Influence of Shield Tunneling on Adjacent Structures and Control Technology

Zhi Ding · Xinjiang Wei · Yong Wu

Influence of Shield Tunneling on Adjacent Structures and Control Technology

中国建筑工业出版社
CHINA ARCHITECTURE & BUILDING PRESS

Springer

Zhi Ding
Department of Civil Engineering
Hangzhou City College
Hangzhou, Zhejiang, China

Xinjiang Wei
Department of Civil Engineering
Hangzhou City College
Hangzhou, Zhejiang, China

Yong Wu
Zhejiang Huadong Mapping
and Engineering Safety Technology
Co. Ltd.,
Hangzhou, Zhejiang, China

ISBN 978-981-19-1133-0 ISBN 978-981-19-1134-7 (eBook)
https://doi.org/10.1007/978-981-19-1134-7

Jointly published with China Architecture & Building Press
The print edition is not for sale in China (Mainland). Customers from China (Mainland) please order the print book from: China Architecture & Building Press.

Translation from the Chinese Simplified language edition: “盾构掘进对邻近建筑物影响及控制技术” by Zhi Ding et al., © China Architecture & Building Press 2018. Published by China Architecture & Building Press.

This Springer imprint is published by the registered company Springer Nature Singapore Pte Ltd.
The registered company address is: 152 Beach Road, #21-01/04 Gateway East, Singapore 189721, Singapore

Preface

Since the establishment of the Beijing Subway preparatory office in 1956, China's subway has gone through more than 60 years of development. According to statistics, by the end of 2016, China had opened and operated 128 rail transit lines, with a total length of 3,832 km. Among them, the shield method has become one of the main construction methods of rail transit construction at home and abroad, especially in the construction of subway tunnels in Hangzhou Bay soft soil area, its application is very wide. Shield tunneling is a typical mechanical process of load variation, which will inevitably produce disturbance and cause soil displacement within the influence range of its construction. If the deformation is too large, it will lead to a series of problems such as cracking and collapse of adjacent structures. For example, the construction of the Shanghai Metro Line 4 led to the collapse of an eight-storey building, and the amount of housing damage insurance caused by tunneling in the UK reaches 400 million pounds (about 4 billion yuan) every year. The tunneling process of the subway shield is actually a dynamic excavation process, which means that before the excavated surface of the tunnel reaches the foundation of the building, the surrounding soil movement caused by shield tunneling has had an impact on the building, which is easily neglected in the past design, construction and research.

To deal with the issue above, our research group focuses on the influence of dynamic tunneling of the shield tunnel. Our research has been supported by the national natural fund project: "saturated soil tunnel excavating area of shallow foundation building foundation, foundation and structure synergy mechanism research" (number: 51508506), the Zhejiang province natural fund project: "the subway train roads long-term settlement of shield tunnel in soft soil under load research" (number: LQ16E080008), Hangzhou major science and technology plan projects: "soft soil iron operating vibration and long-term deformation and mitigation control key technology and application" (number: 20172016A06), etc. To calculate the cases of shield construction, the soil loss caused by shield tunneling is regarded as the combination of two kinds of soil loss: the soil loss caused by excavated surface balance, and the soil loss caused by shield tail clearance and grouting, then the modified Sagaseta ground deformation calculation formula can be given. The influence of longitudinal tunneling of shield tunnel on adjacent structures with shallow foundation and short pile foundation is analyzed by using the synergetic theory model, and the deformation law and internal force distribution law of buildings in shield tunneling area are obtained. To study the law of soil lateral displacement caused by tunneling construction of the shield near buildings, this paper innovatively puts forward that the surface transverse subsidence trough can be characterized by three curves: "plug-shaped distribution curve", "skew-shaped distribution curve", and "normal distribution curve" when the tunnel construction is carried out under three working conditions: tunnel is directly under the building, tunnel is in the disturbed area, and tunnel is outside the disturbed area, respectively. Furthermore, the calculation formula and the related parameters of the "plug-shaped distribution curve" and "skew-shaped distribution curve" are presented.

It can be seen that this book is an in-depth summary of the author's scientific research and engineering practice on the deformation of buildings caused by shield tunneling, aiming to help relevant practitioners understand the mechanism, prediction and analysis methods and corresponding control techniques of the influence of shield tunneling on the deformation of surrounding buildings. There are 8 chapters in this book, including the present situation and progress of the research on the deformation of soil and buildings caused by shield tunneling, measurement and analysis of shield tunneling settlement of adjacent different foundation buildings, calculation of soil deformation caused by shield tunneling, a theoretical study on the influence of longitudinal tunneling of shield on adjacent shallow foundation buildings, theoretical research on the influence of longitudinal tunneling of shield on adjacent short pile foundation buildings, study on the influence and control standard of double line shield tunneling on adjacent buildings, prediction of lateral settlement caused by shield tunneling of adjacent buildings, and the construction control technology of shield construction for adjacent structures.

In the process of writing this book, I have received the guidance, suggestions and help from Prof. Zhangyu Ou, Prof. Tangdai Xia, Prof. Senior Engineer Shaojie Zhu, Prof. Gang Wei, Prof. Shimin Zhang, Senior Engineer Xingfu Yu and Senior Engineer Jianshe Qin, and I would like to express my heartfelt thanks to them! Moreover, special thanks to Xiao Zhang from Hangzhou City College, Dr. Bowen Kong from Zhejiang University, Dr. Shengyi Shi from American Texas A&M University, Juncong Fan from Hongrun Construction Group Co., Ltd., Yesheng Wang, Jianghua Huang and Jianfeng Guo from China Huadong Engineering Co., Ltd. for their hard work in data collection, graphing and theoretical calculation. At the same time, I would like to express my heartfelt thanks to the relevant engineers and cooperative organizations for cooperating with this study.

The division of labor in this book is as follows: Chap. 1 is written by Ding Zhi and Wei Xinjiang; Chaps. 2–7 are written by Ding Zhi. Chapter 8 was written by Wu Yong. This book cites a large number of references, including a variety of academic journals and monographs, but inevitably there will be some omissions, I hereby apologize for your understanding and thank you! Due to the limited level, competence and available information of the author, there are inevitably some inadequacies in the book, and we invite your criticism and correction from experts, peers and readers.

October 2017

Zhi Ding
Hangzhou City College
Hangzhou, Zhejiang, China

Brief Introduction

This book adopts the method of combining theory research, field test and engineering practice to make a systematic analysis of the influence of deformation on surrounding buildings during shield tunneling.

This book is divided into 8 chapters, including: introduction, measurement and analysis of shield tunneling settlement of adjacent different foundation buildings, calculation of soil deformation caused by shield tunneling, a theoretical study on the influence of longitudinal tunneling of shield on adjacent shallow foundation buildings, theoretical research on the influence of longitudinal tunneling of shield on adjacent short pile foundation buildings, study on the influence and control standard of double line shield tunneling on adjacent buildings, prediction of lateral settlement caused by shield tunneling of adjacent buildings, and the construction control technology of shield construction for adjacent structures.

The book structure is rigorous, detail, easy to understand and is equipped with a large number of graphs and the theoretical calculation formula, designed to help readers to the rapid and thorough understanding of shield tunneling in the process of surrounding buildings deformation calculation, prediction and control-related issues, such as cultivating readers to solve the problem of shield tunneling on surrounding environment impact of the basic ability and innovative ability.

This book can be used as a teaching and research reference book for teachers and students in civil engineering, traffic engineering, road engineering and other institutions of higher learning, as well as a reference book for technical personnel engaged in the field of underground and tunnel engineering.

Contents

Author's Brief Introduction

Dr. Zhi Ding is a professor works at Department of Civil Engineering, Hangzhou City College, associate dean of the school of engineering, the head of Civil Engineering, mainly engaged in scientific research on intelligent construction and maintenance of shield tunnel. He is a postgraduate tutor of Zhejiang University, a national registered geotechnical engineer, a scientific and technological worker with outstanding achievements in Hangzhou, a candidate for funding from the Zhejiang Association for Science and Technology, and an outstanding teacher in the Hangzhou Education Bureau system. He has presided over more than 20 scientific research projects, such as general and youth projects of the National Natural Science Foundation of China, key and youth projects of the Natural Science Foundation of Zhejiang Province, and major scientific and technological projects of Hangzhou. He has published more than 70 academic papers as the first or corresponding author, he has been cited more than 1,000 times, and his research results have won more than 10 scientific and technological awards (including 5 provincial or ministerial awards).

Chapter 1
Introduction

1.1 The Introduction

With the rapid development of China's economy, higher requirements have been put forward for the urban traffic service function. Rail transit has become the best mode of transportation for the sustainable development of the city due to its advantages such as provincial footprint and high speed. According to statistics, by 2016, 42 cities in the Chinese mainland, including Beijing, Tianjin, Shanghai, Guangzhou, Nanjing, Shenzhen and Hangzhou, had built or were about to start building subways.

In the construction of subway tunnel in soft soil area, shield tunnel-construction method has become one of the most important construction methods because of its obvious economic and technological advantages and little influence on the surrounding environment [1]. However, no matter which form of shield is used to construct underground tunnels, soil displacement will occur to different degrees. For example, surface settlement caused by tunnelling in some sections of Shenzhen area is up to 500 mm [2]. When the soil displacement reaches a certain degree, it will cause the deformation and cracking of the surface buildings, especially for the shallow foundation buildings and ancient buildings, it is more likely to be destroyed and collapsed, thus causing great losses [3–5]. The construction of Shanghai metro Line 4 resulted in the collapse of the skirt of an 8-storey building, and the neighboring Linjiang Garden building showed a relatively obvious settlement (the maximum settlement exceeded 7 mm within one hour, and the maximum cumulative settlement even reached 15.6 mm) [6]. The amount of home damage insurance in The UK due to tunnelling is 400 million pounds a year, and is increasing every year. For example, the JLE subway project in the UK has spent 100 million pounds on building protection, accounting for one sixth of the total cost of civil engineering [7].

The tunneling process of subway shield is actually a dynamic excavation process, and the deformation trough caused by soil loss and other factors should be a THREE-DIMENSIONAL settlement trough [8]. This means that the surface deformation caused by shield tunneling has already affected the building before the tunnel excavation surface reaches the building foundation, which is easy to be ignored in the

Z. Ding et al., *Influence of Shield Tunneling on Adjacent Structures and Control Technology*, https://doi.org/10.1007/978-981-19-1134-7_1

previous studies. Therefore, if the influence of dynamic tunnelling on adjacent buildings is not taken into account in the design and construction, it is inevitable that the high maintenance cost of the building will be incurred, and even the direct threat to people's lives and property will result in incalculable losses.

As far as Hangzhou is concerned, it is located in the south end of the Beijing-Hangzhou Grand Canal, the south wing of the Yangtze River Delta, the lower reaches of the Qiantang River and the west end of the Hangzhou Bay. It is an important central city in the Yangtze River Delta region and a transportation hub in southeast China. With an area of 16,596 square kilometers, Hangzhou is an important center of political and economic activity in the Yangtze River Delta, and the demand for network rail transit is very strong. Hangzhou Metro Line 1 officially opened on November 24, 2012, becoming the first Metro line in Zhejiang province and the fourth city in eastern China to have its own Metro line. By August 2017, Hangzhou subway had three lines in operation, namely Line 1, Line 2 and Line 4, with 68 stations and a total operating mileage of 93.7 km. As of August 2017, There are 10 subway lines under construction in Hangzhou, including the airport extension of Line 1, the second and third phase of Line 2, the southern section of Line 4, Line 5 and Line 6, as well as the intercity rail transit lines Lin 'an, Fuyang, Shaoxing and Haining, with a total mileage of about 235.7 km, as shown in Fig. 1.1. By 2022, Hangzhou will have 10 urban lines with a total length of 375.6 km and two intercity lines. The shield tube segments are mostly located in the quaternary soft soil layer, which has the characteristics of large natural water content, high compressibility and low bearing capacity, and is prone to produce large deformation under the action of external load.

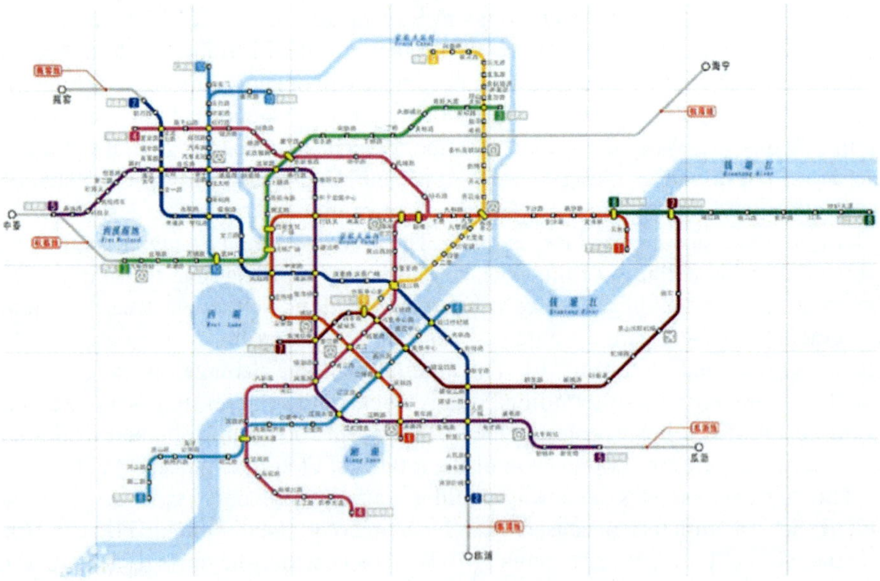

Fig. 1.1 Hangzhou metro line map

Based on the above engineering background, this book takes the shield tunneling with soil pressure balance in soft soil area as the research object, mainly introduces the soil deformation mechanism caused by shield tunneling, and proposes that the soil loss caused by shield tunneling mainly includes the soil loss caused by excavation surface excavation and the soil loss caused by shield tail clearance. Based on the soil loss theory, a mechanical model of the synergistic action of the foundation, foundation and structure above the axis of shield tunnel was established.

1.2 Construction Principle of Earth Pressure Balance Shield Tunneling

Shield tunneling method refers to a tunnel construction method in which the shield tunneling machine controls the excavation surface and surrounding soil not to collapse while tunneling is carried out, the pipe segment lining is assembled inside the shield tunneling machine, and the tail grouting of the shield is carried out simultaneously to prevent excessive soil deformation. Shield tunneling was invented and patented by Brunel, a French engineer, in 1818. The model was an open-type manual tunneling shield tunneling machine, as shown in Fig. 1.2.

Since the 1960s, shield tunneling has been greatly developed and a perfect circular cross section balanced shield tunneling method has been formed, including pressurized air shield, extrusion shield, earth-pressurized shield and mud-water shield, among which earth-pressurized shield and mud-water shield are most commonly used.

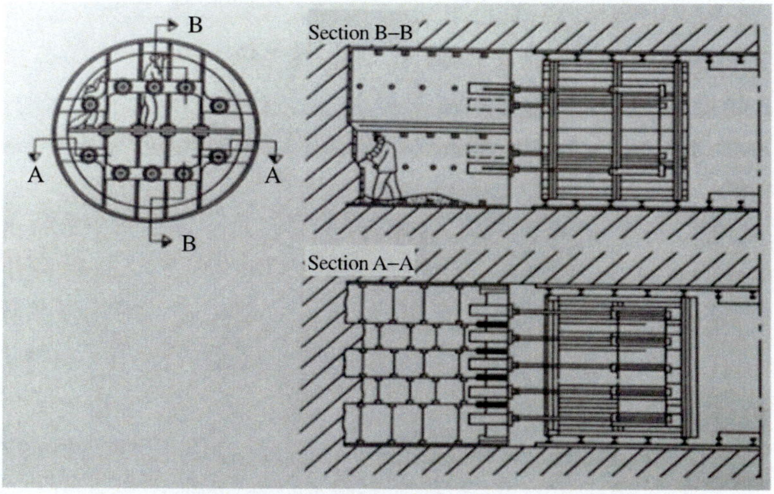

Fig. 1.2 The prototype of the Brunor shield tunneling machine

The earth pressure balance shield was produced by Sato Kogyo in 1963 and was officially used in Tokyo in 1974 for the excavation of a 1.9 km long underground pipeline. The earth pressure balance shield machine uses the full-section rolling cutter head in the front of the shield machine, as shown in Figs 1.3 and 1.4, to cut the soil in front of the shield machine, put it into the sealing cabin behind, and keep the pressure in the cabin balanced with the water and soil pressure of the excavation face, so as to avoid the disturbance of the surrounding soil caused by the shield tunneling, thus reducing the surface settlement. When excavated, the soil is discharged by the screw conveyer at the lower part of the airtight chamber, and a drainage opening is arranged to discharge the soil residue continuously. The output of the spiral conveyor is controlled by the speed of rotation, and the output is closely related to the cutting

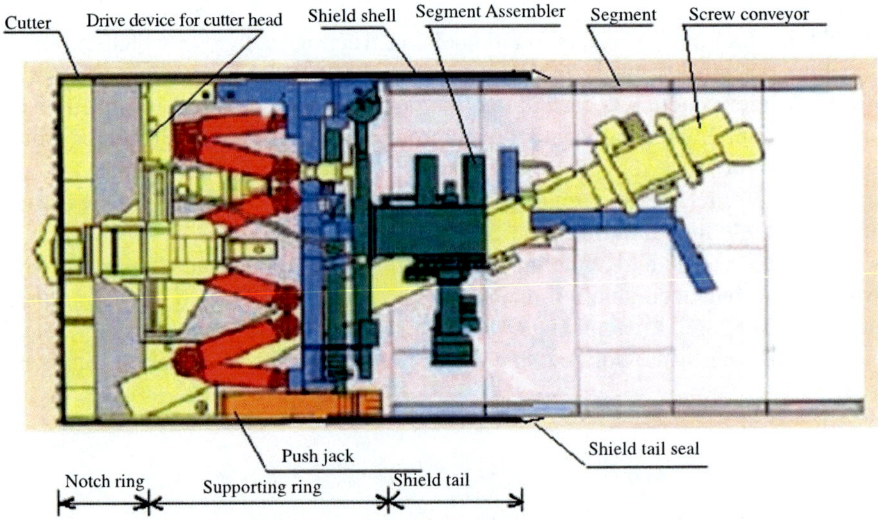

Fig. 1.3 Formation of earth pressure balance shield unit

Fig. 1.4 Factory diagram of earth pressure balancing shield tunnellers

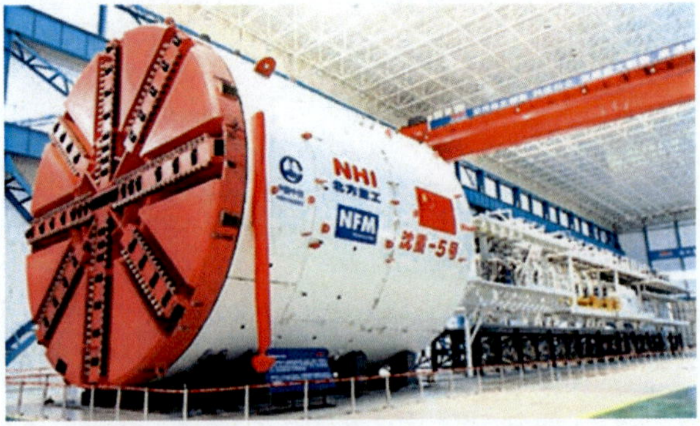

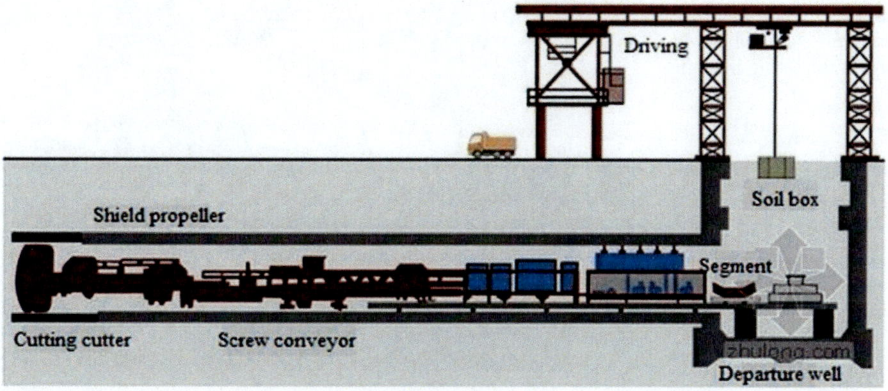

Fig. 1.5 Schematic diagram of shield construction

speed of the cutter, so that the sealed cabin is filled with soil and not too full when cutting.

The construction procedure of the shield tunneling method with earth pressure balance is as follows: excavation, excavation, support and grouting. The construction process is as follows: as shown in Fig. 1.5, the starting shaft or foundation pit is built at one end of the tunnel, and the shield machine is lifted and placed in place. The shield machine starts from the opening of the inner wall of the shaft and advances along the design axis to another shaft. The formation friction resistance encountered during shield tunneling will be transmitted through jacks to the lining pipe segments that have been assembled at the tail part of shield tunneling and then to the back wall of the shaft. For each ring driven by the shield, a ring lining is assembled at the end of the shield, and inert slurry is injected into the gap at the end of the shield in time to prevent excessive ground deformation. The tunneling process ends when the shield tunneling machine reaches the receiving shaft or foundation pit.

In the past more than 30 years, modern shield tunnellers have made great progress in automatic control, hydraulic transmission, back-wall synchronous grouting, pipe piece assembly, computer data acquisition and other aspects. Traditional earth-pressure balanced shield tunnellers have also developed into mud-adding and composite mud-adding earth-pressure balanced shield tunnellers, as shown in Fig. 1.6. The working principle of the mud type earth pressure balance shield machine is to inject plastic material into the sealed bin and fully stir the cut soil at the excavation surface to form a plastic fluid with low permeability. At the same time, the plastic fluid in the bin transmits the set equilibrium pressure to the excavation face, and the servo control is used to match the advancing speed of the shield machine with the earth discharging speed of the screw conveyor, so as to realize the continuous forward driving under the condition of dynamic balance.

When the Japanese Civil Society revised the "Tunnel Standard Specification (Shield) and Interpretation" in 1997, it specially investigated and counted the types

Fig. 1.6 Diagram of a clay earth pressure balancing shield tunneling machine

of shield machines and the number of engineering applications. In the total construction mileage of 811 km, the construction mileage of soil-pressure balanced shield (including mud-adding type and composite mud-adding type) reached 510 km, accounting for 63% of the total mileage. As shown in Table 1.1, the earth-pressure balance shield is widely used in China and can be applied to different geological conditions. It has become the mainstream shield in China's subway construction.

However, whether the traditional or the mud type earth pressure shield is used to construct the subway tunnel, the soil displacement of different degrees will be caused. As early as 1969, Peck pointed out that the soil loss caused by shield construction and its impact on adjacent structures were inseparable from the specific construction details [9]. If the construction control is not proper, the soil around the tunnel will be deformed greatly, and the ground buildings will incline or crack, which will affect the safety and normal use of the surrounding buildings. Therefore, in theoretical analysis, it is necessary to accurately grasp the main influencing factors of shield construction in order to obtain the research results in line with the actual working conditions.

1.3 Research Status of Soil Deformation Caused by Shield Tunneling

The surface deformation curve caused by shield tunneling is generally called "settlement trough", as shown in Fig. 1.7. According to the actual construction conditions in the process of shield tunneling, as shown in Fig. 1.8, the ground deformation caused by shield tunneling can generally be divided into six stages: (1) The soil subsidence or uplift caused by the unbalanced soil and water pressure on the excavation surface during shield tunneling; (2) Friction between shield tunneling machine shell and soil leads to soil uplift; (3) Ground subsidence caused by soil loss caused by shield attitude change; (4) When the shield passes through, the gap in the tail of the shield will cause soil loss and lead to ground subsidence; (5) Ground uplift caused

Table 1.1 List of shield-type metro systems in domestic cities

Place	Formation	Shield machine type	Outer diameter of tube piece (mm)	Outer diameter of tube piece (mm)
Beijing	Clay, silty soil, sandy soil, sand eggs	Soil pressure balance type (including mud type)	6000	5400
Shanghai	Silty clay	Soil pressure balance type, muddy water balance type	6200	5500
Nanjing	Clay, silty soil, sandy soil, highly weathered rock, weathered rock	Soil pressure balance type (including adding mud type), muddy water balance type	6200	5500
Shenzhen	Clay, silty soil, sandy soil, granite, bleach stone, soft above hard below	Soil pressure balance type (including compound mud type)	6700/6000	6000/5400
Dongguan	Soft soil, gneiss, bleach, soft on the bottom hard	Earth pressure equilibrium	6700	6000
Dalian	Bleach, SLATE, diabase	Earth pressure equilibrium	6000	5400
Chengdu	Soft soil, sand egg, stone, mudstone	Soil pressure balance type (also used a mud water balance type)	6000	5400
Fuzhou	Soft soil, moderately weathered granite	Earth pressure equilibrium	6200	5500
Guangzhou	Soft soil, sandy soil, sand eggs, weathered rock, bleach, soft on the bottom hard	Soil pressure balance type (including compound adding mud type), mud-water balance type	6000	5400
Hangzhou	Clay, sandy soil	Earth pressure equilibrium	6200	5500
Kunming	Soft soil, sand egg, mudstone, sandstone	Earth pressure equilibrium	6200	5500

(continued)

Table 1.1 (continued)

Place	Formation	Shield machine type	Outer diameter of tube piece (mm)	Outer diameter of tube piece (mm)
Wuhan	Soft soil, weathered rock	Soil pressure balance type, muddy water balance type	6200/6000	5500/5400
Wuxi	Clay, silty soil	Earth pressure equilibrium	6200	5500
Zhengzhou	Silty clay, sandy soil, sand eggs	Earth pressure equilibrium	6000	5400
Tianjin	Clay, silty, sandy soil	Earth pressure equilibrium	6200	5500
Shenyang	Clay, sandy soil, sand eggs	Earth pressure equilibrium	6000	5400
Suzhou	Silty clay, sandy soil	Earth pressure equilibrium	6200	5500
Xi 'an	Loess, silty soil, sandy soil	Earth pressure equilibrium	6000	5400
Changsha	Silty clay, sand egg, mudstone, sandstone	Earth pressure equilibrium	6000	5400
Nanchang	Silty clay, sandy soil, sandstone and conglomerate	Earth pressure equilibrium	6000	5400
Nanning	Clay, silty soil	Soil pressure balance type, muddy water balance type	6000	5400
Qingdao	Clay, sandy soil, granite	Earth pressure balanced, hard rock boring machine	6000	5400
Hefei	Clay, silty soil	Earth pressure equilibrium	6000	5400
Changchun	Clay, sandy soil, mudstone	Earth pressure equilibrium	6000	5400
Chongqing	Mudstone, granite	Earth pressure balanced, hard rock boring machine	6000	5400
Ningbo	Clay, silty, sandy soil	Earth pressure equilibrium	6200	5500

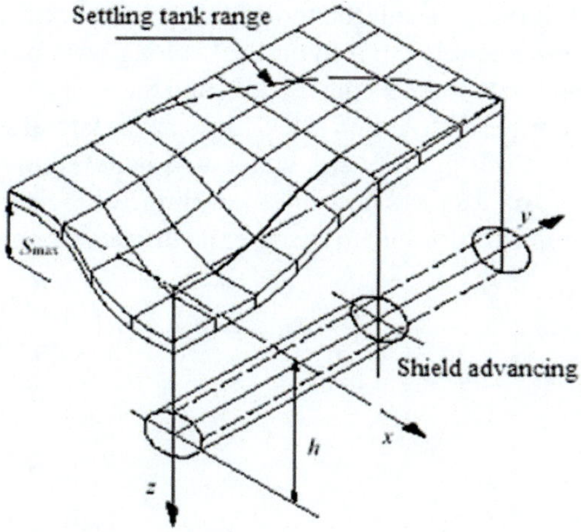

Fig. 1.7 3d settlement trough caused by tunnel excavation

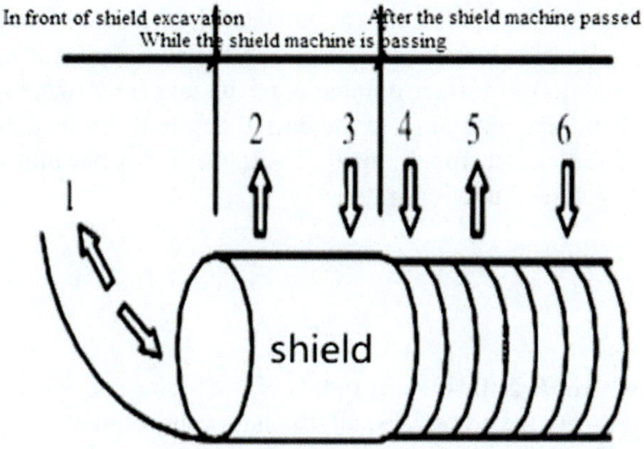

Fig. 1.8 Actual driving conditions of shield tunneling

by grouting behind shield tail wall; (6) Dissipation of excess pore water pressure and consolidation of surrounding soil after shield tunneling lead to ground subsidence [10].

At present, scholars at home and abroad do not consider the consolidation settlement in the 6th stage in the study of soil deformation caused by shield tunneling. The research methods can be summarized as empirical method, theoretical method, numerical method, measurement method and model method.

1.3.1 Empirical Formula Method

(1) Research on lateral surface settlement

The Peck formula is still the most widely used in the empirical prediction of lateral surface settlement [9]. Based on Peck's long-term observation of the surface settlement trough after tunnel construction and the analysis of a large number of measured data, the normal distribution rule of the surface lateral settlement trough during tunnel construction was put forward, which was believed to be caused by soil loss, and the volume of soil loss was equal to the volume of the surface settlement trough.

The empirical formula of its lateral surface settlement distribution is:

$$S(x) = S_{\max} \exp\left(-\frac{x^2}{2i^2}\right) \tag{1.1}$$

$$S_{\max} = \frac{V_{\text{loss}}}{\sqrt{2\pi}i} \tag{1.2}$$

$$V_{\text{loss}} = \eta\pi R^2 \tag{1.3}$$

where: $S_{\max}$ is the maximum lateral surface settlement, mm; $S(x)$ is the surface settlement along the transverse distribution, mm; x is the horizontal distance between the required surface point and the tunnel center line, m; R is the outer diameter of the shield machine, m; h is the buried depth of tunnel axis, m; i is the width coefficient of settlement trough, m; V_{loss} is the soil loss per unit length of shield tunnel, m^3/m; η is the volume loss rate.

$$i = R\left(\frac{h}{2R}\right)^{n} \tag{1.4}$$

where: R is the outside radius of the tunnel, m; h is the depth of tunnel axis, m; $n =$ 0.8–1.0, the softer the soil, the larger the value$n$.

i and V_{loss} in the Peck formula are two important parameters. The correct selection of these two parameters will determine the accuracy of ground settlement prediction. Many scholars have made further studies on the parameter values of the Peck formula: Atkinson and Potts [11], Clough and Schmidt [12], O 'Reilly and News [13], Loganathan and Poulos [14] proposed different methods for the value of i. Cording et al. [15], Attewell [16] has done an in-depth study on V_{loss}.

(2) Study on longitudinal distribution of surface subsidence

The cumulative probability curve formula is still widely used in the empirical prediction of longitudinal surface subsidence [17]. The probability curve formula is proposed to analyze and calculate the longitudinal surface settlement directly above the tunnel:

$$S(y) = S_{\max}\left[\Phi\left(\frac{y - y_{\text{i}}}{i}\right) - \Phi\left(\frac{y - y_{\text{f}}}{i}\right)\right] \tag{1.5}$$

where: $S(y)$—The coordinate position along the direction of tunnel tunneling is the longitudinal surface settlement at y;

y—Coordinates of surface points along the tunneling direction of the tunnel, m;

y_{i}—The starting point of the excavation face heading, m;

y_{f}—The current position of the tunnel excavation face, M;

Φ—This function can be looked up from the standard normal distribution function table.

Liu and Hou [18]. The distribution law of surface settlement caused by tunnel construction in soft land areas such as Shanghai is summarized, and the calculation formula for predicting longitudinal ground settlement is put forward by referring to the formula of Attewell cumulative probability curve:

$$S(y) = \frac{V_{l1}}{\sqrt{2\pi}i}\left[\Phi\left(\frac{y-y_{\mathrm{i}}}{i}\right) - \Phi\left(\frac{y-y_{\mathrm{f}}}{i}\right)\right] + \frac{V_{l2}}{\sqrt{2\pi}i}\left[\Phi\left(\frac{y-y'_{\mathrm{i}}}{i}\right) - \Phi\left(\frac{y-y'_{\mathrm{f}}}{i}\right)\right] \tag{1.6}$$

where: $S(y)$—vertical surface settlement (positive value represents settlement, negative value represents uplift), m;

y—The distance between the settling point and the origin of the coordinate axis, m;

y_{i}—The distance between the starting point of shield tunneling and the origin of the coordinate axis, m;

y_{f}—The distance between the excavation face of shield tunnel and the origin of coordinate axis, m;

L—Shield captain degree, M;

$y'_{\mathrm{i}} = y_{\mathrm{i}} - L; y'_{\mathrm{f}} = y_{\mathrm{f}} - L$;

V_{l1}—Soil loss caused by the excavation face of the shield (the underbreak is negative), m^3/m;

V_{l2}—Soil loss caused by other construction factors such as insufficient grouting volume in the gap at the end of the shield and change of driving direction of the shield tunneling machine after excavation face, m^3/m;

Φ—This function can be looked up from the standard normal distribution function table.

The empirical formula method generally USES Gaussian formula to describe the lateral surface settlement, and accumulative probability curve to describe the longitudinal surface settlement, and determines several key parameters to predict the settlement. Its defect is that it cannot specifically reflect the construction situation, and there are many limitations. However, it is still convenient to predict the surface settlement under certain working conditions, so it is most widely used.

1.3.2 Theoretical Analysis

(1) Huiyuan method

Sagaseta [19, 20] assuming that the soil mass is an incompressible uniform elastic semi-infinite body, and the soil mass loss is equivalent to a cylinder, the three-dimensional surface deformation calculation formula is obtained:

$$\left.\begin{aligned} S_{x0} &= -\frac{V_{loss}}{2\pi}\frac{x}{x^2+h^2}\left[1-\frac{y}{\sqrt{x^2+y^2+h^2}}\right] \\ S_{y0} &= \frac{V_{loss}}{2\pi}\frac{1}{\sqrt{x^2+y^2+h^2}} \\ S_{z0} &= \frac{V_{loss}}{2\pi}\frac{h}{x^2+h^2}\left[1-\frac{y}{\sqrt{x^2+y^2+h^2}}\right] \end{aligned}\right\} \tag{1.7}$$

where: x—the transverse distance from the axis, m;

y—Distance to the excavation face (the heading direction represents the positive direction), m;

H—Depth of tunnel axis, m;

V_{loss}—Formation loss volume, m^3/m.

Verruijt and Booker, based on Sagaseta's "source and sink method" [21], Loganathan [14], Park [22], Chen and Hu [23], Jiang and Zhao [24]. However, the relevant improved formula only considers the soil loss, which cannot reflect the specific influencing factors of construction and the uplift phenomenon at the excavation face.

(2) Mindlin solution

Wei [25], Qi et al. [26], Tang et al. [27], Lin et al. [28]. The elastic Mindlin solution was used to deduce the theoretical formula of formation longitudinal deformation caused by additional thrust of shield and friction between shield and soil, and the deformation caused by soil loss was solved by combining with the mirror image method. However, this kind of theoretical calculation formula cannot well explain the dynamic changes of uplift and settlement at a certain distance in front of shield excavation.

(3) Random medium method

Zhu et al. [29], Shi et al. [30], Qi et al. [26]. The surface deformation caused by tunnel excavation is predicted by using the random medium theory, and the parameter values in the random medium theory are studied, and the calculation formulas of soil deformation caused by different types of shallow buried tunnels are proposed. However, the random medium prediction theory cannot accurately calculate the surface uplift, so most of the studies in the analysis adopt the modified method or negative soil loss to calculate the surface uplift.

1.3.3 Numerical Simulation Method

Shield tunneling is a typical THREE-DIMENSIONAL problem. The soil deformation is closely related to the relative position of the tunneling machine. The numerical simulation method can simulate various construction conditions [31]. A 3-d elastic–plastic finite element model is established to simulate the soil displacement and tunnel surface stress caused by tunnel excavation in soft soil area. The finite element calculation of soil stress field and displacement field under two limit states of non-lining condition and complete lining condition shows that the soil displacement is controlled by the size of the elastic–plastic zone around the excavation surface of the tunnel.

Zhang et al. [32]. More refinement finite element numerical simulation model is put forward, and the party set up in front of the shield excavation excavation unloading unit to simulate the movement of the soil excavation surface, and by applying the known node displacement to simulate shield cutterhead overbreak and shield tail pavement caused by the loss of soil, by setting the vertical and horizontal to the three-dimensional contact element to simulate the moving process of shield machine of relevant contact between soil and structure.

Zhang et al. [33]. Based on the shield tunneling project of Nanjing Metro, a three-dimensional finite element model of shield tunneling, including the dead weight of shield tunneling machine, lining stiffness, thrust of jack and grouting at the tail of shield tunneling machine, was established to further study the surface and tube deformation caused by shield tunneling machine.

Thomas and Gunther [34]. Aiming at the process of earth pressure balance shield tunneling in soft soil area, a three-dimensional model considering many influencing factors of construction is established. The influence of soil pressure balance, grouting at the end of the shield, the weight of the shield tunneling machine and the trailer are considered, and the influence of the movement of the shield tunneling machine and the assembling of lining pipe pieces are analyzed in the simulation process.

Fang and He [35]. Taking the earth-pressure balanced shield machine as the background, the 3d finite element method is used to simulate the construction process of the orthogonal underpass and parallel shield tunnel respectively, and the deformation and internal force variation of the existing tunnel during the dynamic tunneling of the new tunnel are analyzed. In the numerical model, the interaction between the shield tunneling machine and the tunnel segment lining and the transverse isotropic properties of the segment lining structure are considered.

Mroueh and Shahrour [36]. Consideration of shield tunneling construction characteristics, a simplified three-dimensional nonlinear finite element numerical model, the model mainly considered the failed to timely in the process of the excavation support lining length and stress release rate, can think of shield construction process by adjusting the lining support the steps of time and the size of the stress release to control the deformation of the ground.

Zhu et al. [37]. Furthermore, theoretical and empirical formulas were added to the numerical simulation process for prediction, which was firstly based on Rowe et al. [38]. The concept of "gap parameter" is proposed, and the corresponding modification of "gap parameter" is then based on Loganathan formula [14]. Based on the modified "clearance parameters", the construction factors of the surface settlement were quantitatively analyzed, and a numerical simulation model was established for Zhang et al. [39]. Based on the experience, the parameter thickness of equal stratification is modified accordingly.

At present, there are many researches on the surface deformation caused by shield tunneling by finite element method, and many achievements have been made. The finite element method can simulate various working conditions and construction factors, such as the interaction of shield tunneling machine performance, shield tunneling speed and rectifying grouting, etc., with good repeatability. But finite element method is a waste of time, and due to the complexity of the physical and mechanical properties of soil, it is difficult to determine the basic parameters and establish the constitutive model, it is difficult to reflect the actual engineering situation.

1.3.4 Model Test Method

In order to analyze the influencing factors of surface deformation caused by tunnel construction, many scholars have studied this subject through similar material model test, centrifugal model test and other methods. This kind of method is mainly aimed at a specific engineering condition, the model is established in the laboratory according to the similarity theory and the simulation test is carried out, and then the deformation law of surrounding soil in the process of shield tunneling is obtained.

Mair et al. [40]. The ground settlement caused by shallow tunnel was analyzed by centrifuge model test and finite element method. It is believed that the size of land subsidence is largely determined by the characteristics of soil around the tunnel, and the stability coefficient of shallow-buried tunnel is related to the depth of the tunnel itself, and the difference of land subsidence caused by the construction of two tunnels with the same stability rate but different buried depth will be relatively large.

Li et al. [41]. This paper studies the influence of the construction of the shield tunneling under Xuanwu Lake highway tunnel in Nanjing by using the method of indoor model test and finite element, and concludes that the disturbance to the soil around the existing tunnel can be effectively reduced by using appropriate construction parameters.

Liu et al. [42]. By means of large-scale physical model test, the surface subsidence law under different working conditions is summarized, and the surface subsidence curve under different working conditions is summarized, and the influence of the buried depth, support pressure and driving speed of the tunnel on the surface subsidence value is discussed.

He et al. [43]. An indoor similarity model test was carried out for the process of soil pressure balanced shield tunneling, and the general rule of the influence of soil pressure balanced shield tunneling on soil deformation was obtained. The test results could directly reflect the actual formation settlement in the prototype after the conversion of similarity relation.

The advantage of similar model test is that the laboratory experimental conditions can be artificially controlled to study the influence of single or multiple variables on the test results, and the test efficiency is relatively high. In addition, the laboratory model test can be repeated damage experiments, but the site construction process can not meet the corresponding test conditions. However, the similarity model experiment also has its disadvantages, mainly including: the similarity criterion is not easy to be met in general, and the initial and boundary conditions are not easy to be simulated, and the research period and related costs of quantitative analysis test may need to be further increased [44].

1.3.5 Field Measurement Method

Domestic metro field test is early in the 1970s in Shanghai subway tunnel test section, mainly for ground subsidence caused by metro shield construction monitoring, and on the basis of the measured to a certain degree of improvement of Peck formula, proposed under the disturbance of soil consolidation of ground settlement calculation formula.

Zhao [45]. The field measurement of the tunnel from Xinhua Road to Lower tile-house of Tianjin Metro Line 1 was carried out to study the influence of shield construction on surrounding soil, and the deep horizontal displacement, vertical stratified displacement and surface soil deformation were measured respectively, and the displacement variation law of soil mass during shield tunneling was given.

Qu and Xu [46]. The field measurement of the tunnel from Longdong Road Station to Shiji Park Station of Shanghai Metro Line 2 is carried out, the relationship between the buried depth of the tunnel and the maximum surface settlement and the width coefficient of the surface settlement trough is studied, and the measured shapes of the settlement trough under different buried depths are given.

Hu and Huang [47]. A tunnel in Shanghai metro line M4 close wear has been operating under the M2 line engineering construction monitoring, analysis of the shield under the two close in construction of tunnel deformation characteristics, and studied the deformation law of surrounding soils M2 line, think of shield tunneling process must be meet slowly drifting, even turn, pressure maintenance, and proper grouting construction conditions.

Jiang et al. [48]. On guangzhou metro 2, 8, some interval extension field testing was studied in the process of shield tunneling caused by two-phase horizontal displacement of deep, and combining the measured results and the numerical simulation was carried out further calculation, thought in the different stages of shield by, parallel to the horizontal displacement of the tunnel direction and perpendicular to the horizontal displacement of the tunnel direction presents different change law of deep two-phase horizontal displacement are not negligible.

Wei et al. [49]. Of Hangzhou metro line 1 earth pressure balance shield tunnel construction in the field monitoring, the monitoring content includes: the surface settlement, the deep horizontal displacement of soil, the excess pore water pressure and real-time working parameters of shield machine, to study the relationship between the parameters of shield machine's influence on the deformation of soil, think that correct selection of tunneling parameters can effectively maintain the excavation face stability, reduce the deformation of soil.

1.4 Research Status of the Impact of Shield Tunneling on Adjacent Buildings

Shield tunneling will inevitably cause disturbance to the surrounding soil, and when the deformation of the local surface is large, it will often cause settlement, inclination and even cracking of the adjacent buildings, which has become one of the most noteworthy and researched problems in the process of shield tunneling. The surface deformation of buildings was first to investigate the impact of coal mining area, many scholars to the coal mine of the mining area of building deformation and damage of the in-depth study, the mining area of building damage evaluation standard has formed a certain achievement, and the tunnel construction damage evaluation standard also did not establish regional buildings.

At present, a large number of research achievements have been made on the influence of shield tunneling on adjacent buildings at home and abroad. At the present stage, there are three methods for analyzing the influence of shield tunneling on adjacent buildings: theoretical analysis method, numerical simulation method and field measurement method.

1.4.1 Theoretical Analysis

Skepton and MacDonald [50]. Summed up more than ninety related engineering example, determine allowing the differential settlement of ground surface deformation and the total settlement value, and that is a major cause of the cracks of the structures is caused due to the large radius of curvature of the surface settlement curve, but the settlement curve radius of curvature is difficult to measure, and the deformation Angle is relatively easy to measure, so usually can put the measured deformation Angle as the main judgment according to the structure deformation. Burland [51] and Mair et al. [52]. The deflection ratio and horizontal strain are used to define the damage level of a building. Boone [53]. Assuming that the foundation is a flexible foundation consistent with the ground movement, the tensile and shear strains of the frame wall are calculated and analyzed by means of material mechanics and structural mechanics, and the critical tensile and shear strains of the existing structural materials are compared and analyzed, so as to better evaluate the damage of the structure.

Cao [54], Yao [55]. The impact of land settlement on a building is analyzed using the Jubilee MTR Extension line building evaluation standard, but the shear strain and flexural strain are difficult to determine, and the structural characteristics of the building are not considered. Shi et al. [56]. According to the failure Grade of Brick and stone Structures compiled by the Ministry of Coal Industry, the safety of surface buildings in shallow buried tunnel excavation area is evaluated from three aspects: the maximum subsidence value, the tilt value and the bending curvature. Ge et al. [57]. In view of the disturbance caused by shield construction to the building, the dual control indexes are proposed: (1) The average slope of settlement trough under the building is less than 2%; The average slope increment of settlement trough under the building is less than 0.1%.

Richard et al. [58]. The deformation of ground frame structure caused by tunnel excavation is analyzed by analytical method. It is assumed that the ground beam can limit the wall deformation and restrain the shear deformation, and a simplified closed solution is proposed by combining the deflection, shear stiffness and deviation ratio.

Han et al. [59]. The building settlement in THE JLE project of the British subway is fitted, and the Gauss distribution model of multi-storey building settlement with simple shape and uniform structural stiffness is proposed on the basis of Peck formula for calculating the lateral surface settlement. In fact, before the excavation surface reaches the building foundation, the longitudinal surface deformation caused by it has already affected the building.

Zhi et al. [60]. Based on the Boussinesq solution of elastic half space, the influence of surface settlement caused by shield construction of adjacent buildings is analyzed theoretically. Based on the perturbation load, the building load model is introduced to solve the lateral surface settlement caused by shield construction under the condition of adjacent buildings.

Ouyang et al. [61]. Considering the stiffness of the building, the building as the overlying crust layer, on the principle of equivalent stiffness of the building shell layer is analyzed, will cause the deformation problem of shield tunnel through building into a homogeneous semi-infinite space Verruijt and Booker problem solution, thus introduced considering building stiffness of shield tunnel crossing buildings caused by surface transverse settlement calculation formula.

1.4.2 Numerical Simulation Method

The influence factors of shield tunneling on adjacent buildings are relatively complex, and the existing theoretical analysis methods are still insufficient in the study of building deformation and internal force change, so the finite element numerical analysis method is still one of the main research methods for scholars at home and abroad.

Mroueh and Shahrour [62]. The three-dimensional numerical simulation of the surface building deformation caused by tunnel excavation shows that ignoring the dead weight of the building will lead to significantly smaller settlement calculation results, but it does not take into account the characteristics of the building structure, so that the surface settlement will produce significant changes at the independent foundation joints.

Jenck and Dias [63]. The influence of soil loss is considered, and the influence of the change of building stiffness on surface deformation is studied. It is believed that the existence of buildings has obvious influence on the surface settlement and should be paid attention to in the process of shield construction.

Jiang et al. [64]. The dynamic simulation of shield tunneling process was carried out by using finite element program ABAQUS, and the influence of shield tunneling on adjacent buildings was further analyzed. It is believed that the closer the building is to the tunnel, the greater the influence of shield tunneling is. When the distance is beyond 3 times the radius, the influence is already small; when the distance is greater than 4 times the radius, the influence can be ignored.

He et al. [65]. The structural settlement and foundation inclination of the adjacent high-rise buildings caused by the construction of new tunnel are studied by using the finite element method. It is considered that the building slopes away from the tunnel in the adjacent affected area before the shield arrives. In the process of shield tunneling, the building inclines to a certain extent toward the adjacent tunnel until it becomes stable in the later stage.

Ding et al. [66]. A 3d numerical model was established by Midas/GTS finite element software, taking the tunnel and the building into account 90°, 60°, 45° And 30° The influence of tunnel excavation on the foundation settlement and structural deformation of surface buildings under different angles of tunnel and buildings is analyzed under four different working conditions.

Yao and Yang [67]. In Beijing subway line 10 of shield tunnel side wear a raft foundation engineering example of the building, the FLAC3D finite element software is used for three dimensional numerical calculation, studied the shield to buildings, through and left the building before the three stages of the surface vertical and horizontal deformation law, and put forward the building foundation deformation control according to the specification standard.

1.4.3 Field Measurement Method

Breth and Chambosse [68]. The land subsidence caused by shield tunneling under masonry structure and frame structure is analyzed, and it is found that the actual measured land subsidence trough width is much wider than the predicted result without considering the building stiffness, and the stiffness of the building must be considered when predicting the impact of shield tunneling on the building.

Yang et al. [69]. Based on the field monitoring of a brick-concrete house under the shield of Shanghai Rail transit Line M8, it is believed that the soil loss has the greatest influence on the surface settlement, and construction parameters such as tunneling speed, soil bunker pressure and excavated quantity should be reasonably controlled during the process of shield propulsion.

Li [70]. The experiment of suzhou Metro Line 1 running through the canteen of The New District Experimental Middle School is carried out, the key technologies of shield tunneling through independent foundation buildings in suzhou powder land are summarized, and the methods of shield tunneling parameter control, tracking grouting and ground reinforcement protection are put forward systematically.

Sun and Guan [71]. Based on an example of a residential group of masonry structures under shield tunneling in a section of Hangzhou Metro Line 1, the influence law of shield tunneling on the settlement of adjacent masonry structures is studied by monitoring and analyzing the settlement of buildings during the whole construction period of left and right tunnel.

Xu et al. [72]. A range in Tianjin metro line 3 were measured under wear style building, shield tunneling is studied in the process of building deformation regularity and characteristics of think the reasonable control of shield tunneling parameters can effectively reduce the settlement of the building, under the muddy soil grouting can lift the weight of smaller buildings, but the pore pressure dissipation caused by grouting leads to post-construction settlement building.

With the vigorous development of subway construction in China, more and more cases and studies based on field measurement analysis have been conducted. Such studies are of certain engineering significance for guiding the shield construction in the future. However, field monitoring test generally costs a lot of people, materials and financial resources, and may be limited by actual construction conditions, so it cannot be carried out in an all-round way. Therefore, the measured data of structure deformation caused by subway shield construction in different areas need to be further summarized and analyzed.

1.5 Deficiencies of Existing Studies

The above research status at home and abroad indicates that there are still many deficiencies in the calculation and analysis of the impact of shield tunneling on adjacent buildings. Specific problems can be summarized as follows:

(1) the existing calculation and prediction of shield tunnel excavation deformation still is given priority to with empirical formula, the theoretical solution of Mindlin method and random medium method is applied in the engineering field, harder though Sagaseta solution can be applied, but the existing calculation will shield tail and the soil loss caused by the excavation face confusion

together, without considering the construction characteristics and the influence of technological factors, it is necessary for further revision.

(2) Domestic and foreign research methods on the impact of shield tunneling on buildings are still mainly focused on numerical simulation methods. There are mainly two types of analysis methods. The first type is the holistic analysis method, which does not consider the impact of buildings on soil deformation caused by shield tunneling. The second type is two-stage analysis, but the type, basic form and stress characteristics of the building are not considered, so the applicability of the research results is greatly reduced.

(3) the achievements are concentrated in the lateral surface subsidence caused by tunnel excavation face on study the influence of building deformation, and is simplified as a static disturbance, and consider the subway shield tunnel longitudinal excavation little influence on adjacent shallow foundation building, at the same time, building additional deformation due to tunnel excavation and the additional internal force calculation formula has been reported, reflect the dynamic area of tunnel excavating foundation, shallow foundation analytical solutions and building synergy model also has not yet been established.

(4) The existing formula for predicting the settlement of tunnel excavation face of adjacent buildings is still the Peck formula, but it does not consider the existence of buildings. Most scholars still use Peck's calculation formula for comparative analysis in their studies on the interaction of tunnel-build-soil. But in practical engineering, the deformation curve of the building is completely different from that of the natural homosexual love because of its structural rigidity. Ignoring the dead weight of the building will lead to significant changes in the calculation results of surface settlement and the width of settlement trough.

(5) At present our country building evaluation standard is almost blank in a subway tunnel construction, difficult to determine the effective maximum ground surface settlement of the excavation face of shield tunnel on the adjacent buildings, to determine the damage of the buildings, not to the damage of the buildings and the combination of shield tunneling of soil loss, and can't consider building foundation forms, not better guide the subway engineering practice.

1.6 Main Research Contents of the Book

According to the shortcomings of existing studies at home and abroad, this book mainly focuses on the soil deformation caused by shield tunneling and its impact on different adjacent foundation buildings, and carries out the following researches:

(1) Field measurements were made on the tunnel project between Ganshanmen Station (now renamed As Ditianguan Station) and Zhanongkou Station of Hangzhou Metro Line 1, and the influence of shield tunneling on soil deformation and adjacent different foundation buildings was studied. Focus on actual wear short pile foundation under the shield multi-storey buildings (the first

car company coach brigade office building), the deformation of and adjacent shallow foundation multi-storey buildings (Wen Hui) next to the farmers market building and long pile foundation in high-rise buildings (wild wind modern home residential buildings) deformation caused by shield tunneling, and the construction parameters are analyzed.

(2) Combined with the actual construction of shield, the soil loss caused by shield tunneling is regarded as the soil loss caused by excavated surface balance and the soil loss caused by shield tail clearance and grouting. Based on Sagaseta huiyuan method theory, assuming the excavation face unearthed balance caused by soil loss is related to the unearthed rate caused by the amount of the radial displacement of soil, shield tail gap produced by the soil loss is related by the shield tail escapes and grouting amount of radial displacement of soil, semi-infinite space is deduced formula of three-dimensional deformation of soil, and presents a modified Sagaseta ground deformation calculation formula.

(3) Introducing the theory of synergy model analysis of shield tunnel longitudinal excavation of shallow foundation and short pile foundation adjacent buildings, the influence of the soil loss caused by shield construction as the main cause of land subsidence, and use the modified Sagaseta formula of ground deformation, deduce the shallow foundation building foundation with short pile foundation, foundation and structure of the mechanical model of synergy and the theoretical solution, and use the analysis software 1 stop numerical integral solving, shield tunneling can be gained by analyzing deformation and internal force distribution of area buildings.

(4) Analysis of adjacent building caused the soil horizontal displacement law of shield tunnel construction, put forward the tunnel within the scope of the building directly, disturbance and outside disturbance range in construction of three kinds of working conditions, the surface of the horizontal sedimentation tank is respectively "plug shape distribution curve", "skewness distribution curve", and "normal distribution curve" characteristics, and presents a plug shape distribution curve calculation formula and the formula of skewness distribution curve and related parameters.

(5) On the basis of soil loss calculation theory, the mechanical model of foundation, foundation and structure synergy of buildings above the axis of double-line parallel (double-circle) shield tunnel was further established, and the deformation and internal force variation rules of adjacent shallow foundation buildings during the tunneling process of double-line parallel (double-circle) shield tunnel were analyzed. Based on the above theoretical model and combined with the principle of safety evaluation of building deformation, the paper puts forward the standard of building deformation and safety control in shield tunneling area.

(6) The use of common software development tools Delphi7.0, visualization software is compiled "adjacent building tunnel construction system", can be more convenient to determine the earth's surface caused by tunnel excavation horizontal sedimentation tank size, and whether its damage to nearby buildings produce different distance, which can be more convenient to determine the

degree of structural damage. The ground surface deformation datum value is calculated by the allowable tensile strain and tilt rate of the building, and then the ground surface deformation datum value is used as the evaluation basis to determine whether the adjacent buildings at different distances suffer construction damage.

(7) With three different, Hangzhou, nanchang, ningbo shield tunnel adjacent building, construction of project cases, given the right adjacent buildings of shield construction control technology, proposed the shield machine, buildings and shield tunnel grouting reinforcement, formation, construction monitoring analysis specific technology implementation measures, such as the construction process is analyzed and the surface and the settlement of the building, to judge the rationality and validity of the engineering control technology measures, can provide useful reference for similar projects.

References

1. Mengshu W. An overview of development of railways, tunnels and underground works in China. Tunnel Constr. 2010; 30(4):351–364.
2. Zhang D, Huang J. Analysis of ground deformation for shallow tunnel in Shenzhen city. J China Univ Min Technol. 2004; 33(5): 578–583.
3. Ma J. Settlement control measures for long distance shied excavating beneath residential areas. Munic Eng Technol. 2011; 29:90–92.
4. Zhonghua W. Construction technology of shield machine development in unfavorable stratum under the existing building. Eng Constr. 2011; 43(2):33–37.
5. Hou C, Kang Z, Hou X et al. Analysis of ground subsidence of south moat section in the construction of Xi'an No.2 subway line. J Hefei Univ Technol (Natural Science). 2011; 34(1):102–104.
6. Wang X, Sun B, Yu C. Influence of subway construction on urban environment and its countermeasures. Urban Mass Transit. 2004; 3:8–89.
7. Han X. The analysis and prediction of tunnelling-induced building deformations. Xi'an: Xi'an University of Technology;2006.
8. Attewell PB, Yeates J, Selby AR. Soil movements induced by tunneling and their effects on pipelines and structures. Glasgow: Chapman & Hall; 1986.
9. Peck RB. Deep excavations and tunneling in soft ground. In: Proceeding of 7th international conference on soil mechanics and foundation engineering. Mexico City: State of the Art Report. 1969. p. 225–290.
10. Lin G. Shield construction control technology of tunnel engineering passing through different buildings. China Munic Eng. 2013; 164(1):48–53.
11. Atkinson JH, Potts DM. Subsidence above shallow tunnels in soft ground. J Geotech Eng. 1977;103(4):307–25.
12. Clough GW, Schmidt B. Design and performance of excavations and tunnels in soft clay. New York: Elsevier Science Publishing Company; 1981. p. 569–634.
13. O'Reilly MP, New BM. Settlements above tunnels in the United Kingdom–their magnitude and prediction. In Proceedings of tunnelling'82 symposium. London: Institution of Mining and Metallurgy;1982. p. 173–181.
14. Loganathan N, Poulos HG. Analytical prediction for tunneling-induced ground movement in clays. J Geotech Geoenviron Eng. 1998;124(9):846–56.

15. Cording EJ, Hansmire WH, Macpherson HH, et al. Displacement around tunnels in Soils. Report prepared for department of transportation. Urbana: University of Illinois; 1976.
16. Attewell PB. Ground movements caused by tunneling in soil. In Conference on large ground movements and structures, Cardiff. London: Pentech Press; 1978. p. 812–948.
17. Attewell PB, Woodman JP. Predicting the dynamics of ground settlement and its derivatives caused by tunneling in soil. Ground Eng. 1982; 15(8):13–20, 36.
18. Liu J, Hou X. Shield tunneling method. Beijing: China Railway Publishing House;1991.
19. Sagaseta C. Analysis of undrained soil deformation due to ground loss. Geotechnique. 1987;37(3):301–20.
20. Sagaseta C. Author's reply to Schmidt. Geotechnique. 1988;38(4):647–9.
21. Verruijt A, Booker JR. Surface settlements due to deformation of a tunnel in an elastic half plane. Geotechnique. 1996;46(4):753–6.
22. Park KH. Elastic solution for tunneling-induced ground movements in clays. Int J Geomech. 2004;4(4):310–8.
23. Chen F, Hu Z. Analytical prediction of tunneling induced surface movements due to shield-deviation in undrained soil. Rock Soil Mech. 2004; 25(9):1427–1431.
24. Jiang X, Zhao Z. Application of image method in calculating tunneling-induced soil displacement. J Harbin Inst Technol. 2005; 37(6):801–803.
25. Wei G. Theoretical study on properties of soil and structure during pipe jacking construction. Hangzhou: Zhejiang University;2005.
26. Qi J, Xu R, Wei G. Research on calculation method of soil 3D displacement due to shield tunnel construction. Rock Soil Mech. 2009; 30(8):2442–2446.
27. Tang X, Zhu J, Liu W et al. Research on soil deformation during shield construction process. Chin J Rock Mech Eng. 2010; 29(2):417–422.
28. Lin Z, Zhang Z, Wu S et al. Study of ground heave and subsidence induced by shield tunnelling in soft ground. Chin J Rock Mech Eng. 2011; 30(12):2583–2591.
29. Zhu Z, Zhang Q, Yi H. Stochastic theory for predicting longitudinal settlement in soft-soil tunnel. Rock Soil Mech. 2001; 22(1):56–59.
30. Shi C, Liu B, Chen L. Prediction of longitudinal movement and deformation of stratum in longitudinalsection due to tunnel construction by shield. Chin J Geotech Eng. 2003; 25(5):585–589.
31. Lee KM, Rowe RK. Finite element modeling of the three-dimensional ground deformations due to tunneling in soft cohesive soils: Part I–method of analysis. Comput Geotech. 1990;10:87–109.
32. Zhang H, Yin Z, Zhu J. Elaborate simulation of shield tunneling. Rock Soil Mechanics. 2004; 25(S2):280–284.
33. Zhang Z, He C, She C. Three dimensional finite element modeling of excavation and advancement processes of shield tunnel construction in Nanjing metro. J China Railw Soc. 2005; 27(1):84–89.
34. Thomas K, Gunther M. On the influence of face pressure, grouting pressure and TBM design in soft ground tunnelling. Tunn Undergr Space Technol. 2006;21:160–71.
35. Fang Y, He C. Analysis of influence of undercrossing subway shield tunneling construction on the overlving tunnel. J China Railw Soc. 2007; 29(2):83–88.
36. Mroueh H, Shahrour I. A simplified 3D model for tunnel construction using tunnel boring machines. Tunn Undergr Space Technol. 2008;23:38–45.
37. Zhu C, Li N, Liu H et al. Analysis of ground settlement induced by workmanship of shield tunnelling. Rock Soil Mech. 2011; 32(1):158–164.
38. Rowe RK, Lo KY, Kack GJ. A method of estimating surface settlement above tunnels constructed in soft ground. Can Geotech J. 1983;20(8):11–22.
39. Zhang Y, Yin Z, Xu Y. Analysis of ground deformation caused by shield tunnel. Chin J Rock Mech Eng. 2002; 21(3):388–392.
40. Mair RJ, Gunn MJ, Oreilly MP. Ground movement around shallow tunnels in soft clay. Tunn Tunn. 1983;14(5):45–8.

41. Li W, He C, Zhang Z. ModelTest of constructing shield tunnel under large underground structure. J Southwest Jiaotong Univ. 2005; 40(4):478–483.
42. Liu J, Liu B, Zhang H. Experimental study on the shield tunneling-induced surface settlement. Ind Const. 2011; 41(3):91–98.
43. He C, Wang Y et al. Similarity model test of Earth-Pressure-Balanced shield tunneling process. China Civ Eng J. 2012; 45(2):162–169.
44. Zhou R. Analysis of construction influence of Double Line Shield Tunnels passing through existing building in soft soil. Beijing: Beijing Jiaotong University;2012.
45. Zhao Z. Study of image theory and application of regression method on tunnelling induced soil displacements and stresses. Tianjin: Tianjin University;2004.
46. Qu J, Xu Y. Analysis of ground traverse settlement trough caused by shield construction. Rock Soil Mech. 2006; 27(2):313–322.
47. Hu Q, Huang H. Analysis and monitoring on shield tunneling under existing adjacent tunnel. Chin J Geotech Eng. 2006; 28(1):42–47.
48. Jiang X, Li L, Yuan J et al. Dynamic analysis of strata horizontal displacements induced byshield construction of deep tunnel. Rock Soil Mech. 2011; 32(4):1186–1192.
49. Wei X, Zhou Y, Wei G. Research of EPB shield tunneling parameter relations and their influence on stratum displacement. Rock Soil Mech. 2013; 34(1):73–79.
50. Skepton AW, MacDonald DH. Allowable settlement of buildings. Proc Inst Civ Eng. 1956;13(6):19–32.
51. Burland JB. Assessment of risk of damage to building due to tunnelling and excavation. In: Invited special lecture to IS-Tokyo'95: 1st international conference on earthquake geotechnical engineering;1995.
52. Mair RJ, Tylor RN, Burland JB. Prediction of gound movements and assessment of risk building damage due to bores tunneling. In: Proceedings geotechnical aspect of underground construction in soil ground. Balkema: Rotterdam;1996.
53. Boone SJ. Ground-movement related building damage. J Geotech Eng. 1996;11:886–896.
54. Cao H. Analysis on the influence of a pipeline tunnel project on the underpass building. Railw Surv Des. 2005; 4:32–34.
55. Yao H, Wang M, Zhang D et al. Security aspect and measurements while tunneling of the thermalpower tunnel from below buildings on ground surface. Rock Soil Mech. 2006; 27(1):112–116.
56. Shi C, Peng L, Liu B. Influence of shallow tunnel excavation on ground surface buildings. Chin J Rock Mech Eng. 2004; 23(19):3310–3316.
57. Ge S, Xie D, Ding W et al. Undercrossing disturbance control criterion for shield tunnel with consideration of building existing deformation. J Tongji Univ (Natural Science). 2011; 39(11):1616–1621.
58. Richard J, Finno FT et al. Evaluating damage potential in buildings affected by excavations. J Geotech Geoenviron Eng. 2005; 131(10):1199–1210.
59. Han X, Standing JR, Li N. Modified stiffiness approach to predict deformation of building induced by tunnelling. Chin J Geotech Eng. 2009; 31(4):539–545.
60. Zhi D, Xinjiang W, Tao Z, et al. Analysis and discussion on surface settlement induced by shield tunnel construction of adjacent structure. Disaster Adv. 2012;5(4):1656–60.
61. Ouyang W, Xie D, Ding W. Calculation method for settlement due to shield tunnelling considering structure stiffness. Chin J Undergr Space Eng. 2013; 9(1):155–160.
62. Mroueh H, Shahrour I. A full 3-D finite element analysis of tunneling-adjacent structures interaction. Comput Geotech. 2003;30:245–53.
63. Jenck O, Dias D. 3D-finite difference analysis of the interaction between concrete building and shallow tunneling. Geotechnique. 2004;54(8):519–28.
64. Jiang X, Zhao Z, Li L. Dynamic simulation of the effects of shield tunneling in Tianjin subway project on neighboring job facilities. J Tianjin Univ (Science and Technology). 2006; 39(2):188–193.
65. He M, Liu J, Le G. Study of impact of shield tunneling side-crossing on adjacent high buildings. Chin J Rock Mech Eng. 2010; 29(3):603–608.

66. Ding Z, Peng L, Shi C. Analysis of influence of metro tunnel crossing angles on ground buildings. Rock Soil Mech. 2011; 32(11):3387–3392
67. Yao A, Yang X. Deformation response and safety evaluation of side-crossing raft slab foundation in shield tunnel construction. Chin J Undergr Space Eng. 2012; 8(4):842–846.
68. Breth H, Chambosse G. Settlement behavior of buildings above subway tunnels in Frankfurt clay. In: Proceedings of the conference on settlement of structures. London, England: Pentech Press;1974.
69. Yang X, Mei Y, Zheng S. Feasibility analysis and monitoring of subway passing through pile foundation of existing buildings. Build Constr. 2006; 28(6):412–415.
70. Li H. Control technology and monitoring of shield tunnel under buildings. Railw Eng. 2011; 9:66–68.
71. Sun Y, Guan F. Shield tunnel construction induced influence on the settlement of masonry buildings. China Railw Sci. 2012; 33(4):38–44.
72. Xu Z, Han Q, Zheng G. Field monitoring and analysis of effects of metro tunnels underhistoric buildings. Chin J Geotech Eng. 2013; 35(2):364–374.

Chapter 2
Measurement and Analysis of Shield Tunneling Settlement of Adjacent Different Foundation Buildings

2.1 The Introduction

China's coastal areas are economically developed and densely populated, where subway construction is concentrated, and where deep soft clay is widely distributed. Subway tunnels inevitably need to cross the weak soil layer with low strength, high compressibility and high sensitivity. At the same time, the subway construction generally goes through the busy commercial downtown, different forms of foundation buildings and underground pipelines are dense. Therefore, how to accurately predict the impact of tunnel tunneling on different adjacent infrastructure buildings and how to reasonably control and reduce the impact has become a major problem to be solved urgently in the construction of metro in soft soil area.

It can be seen from the above review that field test has become one of the effective means to study the soil and building deformation caused by shield tunneling. Foreign countries such as Germany, The United Kingdom, France and Japan have carried out a large number of field tests and theoretical studies in combination with shield tunnel projects. In China, Shanghai, Beijing, Guangdong, Jiangsu and other places have carried out a series of field monitoring studies in combination with shield tunnel projects. However, due to the complexity of engineering geology and the variation of construction parameters, the deformation of soil and buildings in different areas has different characteristics, and the research results are difficult to meet the needs of complex subway shield tunnel construction.

At present, the systematic analysis of the impact of shield tunneling on the deformation of different foundation buildings is rare, and even less so for Hangzhou, a special soft land area. Considering the regional nature of geotechnical engineering, it is necessary to fully discuss and analyze the influence of shield tunneling on the deformation of different foundation buildings, so as to provide reference for similar projects. Based on a large number of monitoring data of different foundation buildings in the Tunnel project of No.1 Subway line in Hangzhou, this chapter preliminarily discusses and analyzes the soil mass and the deformation law of different foundation buildings in the shield tunneling area. On the basis of the measured data, considering

Z. Ding et al., *Influence of Shield Tunneling on Adjacent Structures and Control Technology*, https://doi.org/10.1007/978-981-19-1134-7_2

the influences of construction parameters is deduced from the theoretical solution of deformation of soil (Chap. 3), and deformation in the soil solution is established on the basis of the tunnel—foundation—different basic theoretical model (Chap. 4, 5, 6), the shield tunnel longitudinal excavation on nearby the influence of different foundation building deformation and internal force. In addition, the lateral surface settlement distribution of adjacent shallow foundation buildings in tunneling area is predicted, and the corresponding prediction software is developed according to the existing control standards (Chap. 7).

2.2 Project Overview

2.2.1 Project Introduction

The study object is the tunnel project from Hangzhou metro line 1 gen station (now renamed Datieguan station) to the Zhalongkou station, which is the underground dual single round shield tunnel (shield 5, 6), shield 5, 6, respectively, starting on June 20, 2009, and October 24, and arriving on December 9, 2009 and May 20, 2010. The starting point is Jiaojiacun station, passing through Shaoxing Road, entering railway Ganshanmen Danyun station at Pile Number K18 + 140, connecting with the east of Wenhui Bridge at Pile number K18 + 600, and following Wenhui Road to Zhaliangkou station, the subway interval passing through Shaoxing Road is about 40.0 m wide and Wenhui Road is about 50.0 m wide. Railway Freight station is now existing part of the construction and dense railway lines.

The tunneling machine is "Xizi" (As shown in Fig. 2.1), the first shield tunneling machine made by Hangzhou Boiler Group. The length is 8.5 m, the diameter is 6.34 m, the total length is 60 m and the weight is 350 tons. The lining type is in the form of standard ring plus left turning ring or right turning ring, which adopts the

Fig. 2.1 "Xizi" shield machine

staggered assembly form. The inner diameter of the standard ring pipe is 5.5 m, the outer diameter is 6.2 m, the thickness is 0.35 m and the width is 1.2 m. The reinforced concrete pipe is made of C50 waterproof concrete pouring, and the anti-permeability grade is S10. Each ring is composed of 6 pipe pieces in total, including 1 top sealing block, 2 standard blocks, 2 adjacent blocks and 1 bottom sealing block.

2.2.2 *Hydrogeological Conditions*

(1) Terrain and landform

The project of Ganshanmen Station ~ Zhalongkou Station of Metro Line 1 is located in the east of Hangzhou city and the north of Qiantang River, belonging to the geomorphic unit of alluvial plain. The route of the tunnel is east–west, the terrain of the project site area is flat, the natural ground of the proposed site is relatively flat, and the elevation of the ground is 5.0–6.5 m.

(2) Geological conditions

Within 20 m of the depth of the on-site construction site, there is a fluid-plastic silty clay layer with a thickness of 10.0–20.0 m buried 20–40 m deep, a silty clay layer with a thickness of 1.0–6.0 m buried and a silty clay layer with a thickness of 40–45 m buried, and a round gravel layer with a thickness of more than 3 m at the bottom. The site grade is level 2, and it is medium complex site.

According to the depth of exploration, the site can be divided into ①, ③, ④, ⑥ and ⑧, and is divided into 8 large layers, which can be divided into 18 sub-layers. The physical and mechanical properties of some soil layers are shown in Table 2.1.

(3) Groundwater

The main type of groundwater in this site is quaternary unconsolidated rock pore water, which can be divided into pore water and confined water according to its water-bearing medium, hydrologic property, occurrence conditions and hydraulic characteristics.

1) Diving

The shallow groundwater of this site belongs to pore diving, mainly occurs in the surface fill and ③$_2$~③$_6$ of Stratified silty soil and silty sand, which are replenished by surface water runoff and meteoric precipitation, and the river water is replenished in the area near Maimiao port. The groundwater level changes with the seasons. The buried depth of the borehole static water level is 0.9–3.0 m, and the corresponding elevation is 3.46–5.30 m. According to the hydrogeological data in this region, the annual variation of the shallow groundwater level is about 1.0–2.0 m, and the annual average buried depth of the high water level is about 0.5–1.0 m. According to the experience of similar projects and site environment in Hangzhou urban area, the groundwater flow rate is low.

Table 2.1 Physical and mechanical properties of soil layer

Layer no	Name of the soil	Natural gravity	The compression modulus	Cohesive force	In the Angle of	The permeability coefficient	
		$\gamma(kN/m^3)$	$E_{S1\text{-}2}$(MPa)	c (kPa)	$\varphi(°)$	K_v(cm/s)	K_h(cm/s)
①$_1$	Miscellaneous fill	17.8	–	–	–	3.5×10^{-3}	–
①$_2$	Grain filling	19.2	6.50	28.0	19.5	4×10^{-4}	–
③$_2$	Sandy silt	19.0	12.22	9.4	29.3	3.34×10^{-5}	2.70×10^{-5}
③$_3$	Sandy silt	19.1	13.28	5.7	32.3	7.50×10^{-4}	–
③$_5$	Sandy silt	18.4	9.60	6.0	29.5	1.85×10^{-4}	4.42×10^{-5}
④$_2$	Muddy soil with muddy clay	17.1	2.49	14.9	8.9	5.25×10^{-7}	2.38×10^{-6}
④$_3$	Silty clay	16.9	2.48	15.5	9.4	3.28×10^{-7}	6.28×10^{-7}
⑥$_1$	Silty soil with silty clay	17.8	3.06	17.9	11.0	5.77×10^{-7}	6.98×10^{-6}
⑥$_2$	Silty clay	17.2	3.00	16.2	10.4	9.37×10^{-7}	1.15×10^{-6}
⑧$_1$	Silty clay	17.1	2.80	19.0	8.8	2.62×10^{-7}	4.07×10^{-7}

2) Confined water

In the engineering area, the pore confined water is mainly distributed in the deep ⑭$_2$ layer set of boulder, the amount of water is relatively rich, the water separation layer is the upper sludge clay layer (No.④, ⑥, ⑧, ⑨ and ⑬).

Iron bushing shall be buried in drill hole Z-12 at the position of side passage to isolate the upper aquifer $_2$. The water head buried deep in the circular gravel layer, and the test results of each hole are shown in Table 2.2.

Table 2.2 Test results of artesian head

Test hole	Location	Mileage pile number	Buried depth of artesian water	Corresponding elevation
C-18	Gen gate station	K17 + 780.00	9.51	4.25
Z-06	ZhaLongKou station	K19 + 470.00	9.10	4.00
Z-12	Gen Shan Gate station ~ Zhaolongkou Station	K18 + 725.00	9.78	4.04

2.3 Field Test Plan

Shield construction is a process of continuous disturbance of soil, the soil is squeezed or loss of soil and soil consolidation will cause the ground to produce heavy change, which is related to the following factors: shield equilibrium pressure seal storehouse, the unearthed speed, shield posture, shield shell towing function, degree of segment lining joint seal clearance, the deformation of tunnel lining, consolidation and secondary consolidation settlement of soil solidification shrinkage settlement, grouting filling materials, etc. In order to reduce the adverse impact on the environment, information-based monitoring means must be introduced in shield construction to guide the construction with feedback, ensure the stability of excavation face, correctly control the tunneling speed, and constantly optimize the tunneling construction parameters, so as to effectively control the settlement and deformation of soil.

This chapter focuses on studying the deformation law of the surface and buildings in the area affected by shield tunneling, so it focuses on analyzing the vertical displacement monitoring results of the surface and buildings in the engineering construction.

2.3.1 Test Method

In order to improve the accuracy of monitoring data, the principle of overall and hierarchical network layout is adopted: first, the unified monitoring control network is arranged, and then the monitoring points are arranged on the corresponding basis.

(1) Settlement deformation monitoring elevation control network measurement

1) Independent leveling system shall be adopted, and a set of firm leveling points shall be arranged on both sides far away from certain influence range of construction, which shall be permanent leveling reference points. The datum network for settlement deformation monitoring takes the datum point as the corresponding starting point and forms the horizontal control network for vertical displacement joint measurement.

As shown in Fig. 2.2, the base point is the reference point for observing the vertical displacement of the settlement point. Generally, high-precision leveling method is adopted to measure the elevation of the corresponding base point, and check and verify whether the elevation is changed due to external influences. Back and forth observation with corresponding second-class leveling point of the country, and the inspection period is less than 30 days. At the same time, during the settlement observation, the apparent distance between each measuring point (as shown in Figs. 2.3 and 2.4) and the corresponding rear apparent base point should be controlled, and the apparent difference between them should be less than 2 m. After the observation of each settlement point, the corresponding rear apparent base point must be observed

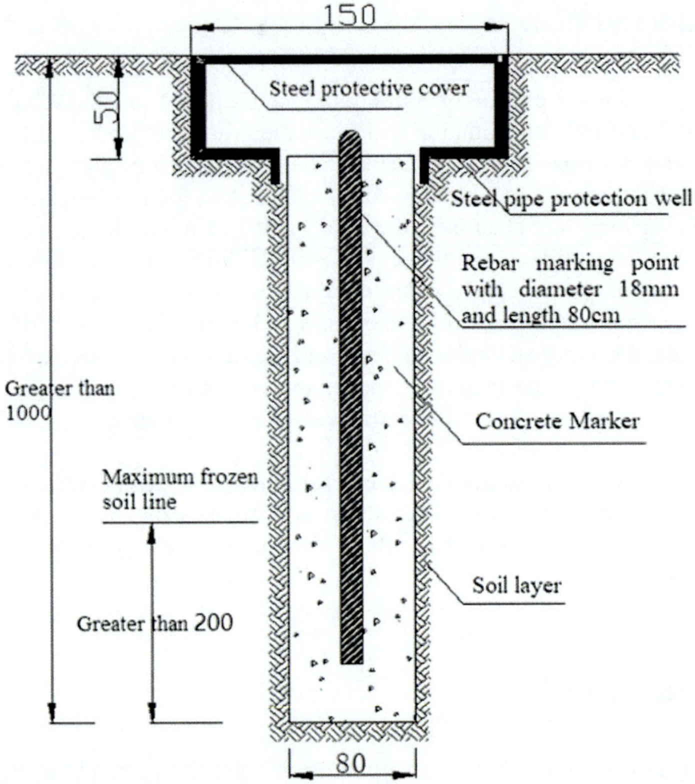

Fig. 2.2 Embedment form of surface reference mark (mm)

again. The difference between the two rear apparent readings should be less than 0.1 mm; otherwise, retest should be conducted.

The observation of the reference network shall be measured in accordance with the national second-class leveling standards. The main technical requirements of high-precision leveling are as shown in Table 2.3.

2) The datum network for elevation monitoring USES WILD NA2 automatic leveling level (nominal accuracy: ± 0.3 mm/km) and supporting indium steel ruler. See Fig. 2.5a, b for the instrument.

The settlement observation should be carried out strictly in accordance with the requirements of the national second-class precision leveling standard. In order to ensure a high observation accuracy, the following measures should be taken during the measurement:

i. Before measurement, the corresponding operation planning table should be prepared to ensure the orderly development of settlement observation;
ii. Comprehensive inspection of high precision level and indium steel ruler should be conducted before observation;

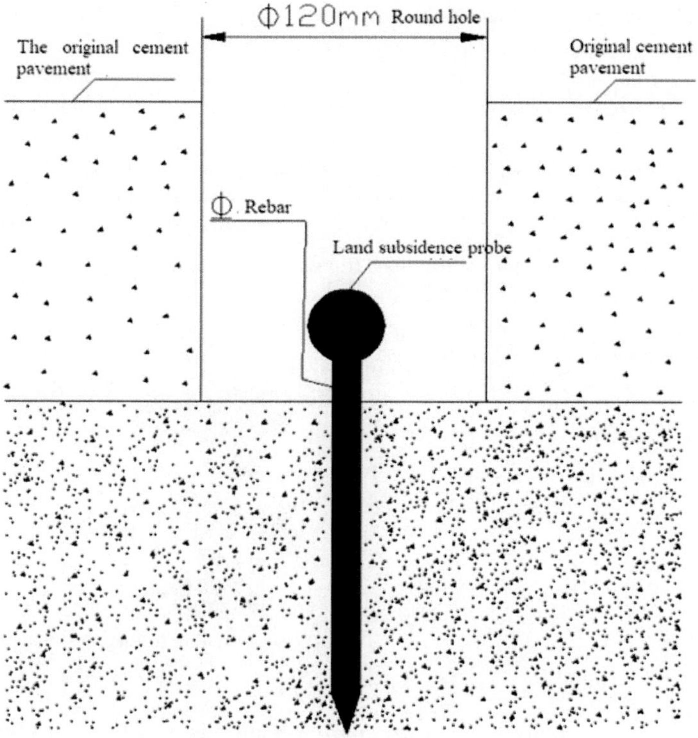

Fig. 2.3 Installation diagram of ground settlement points

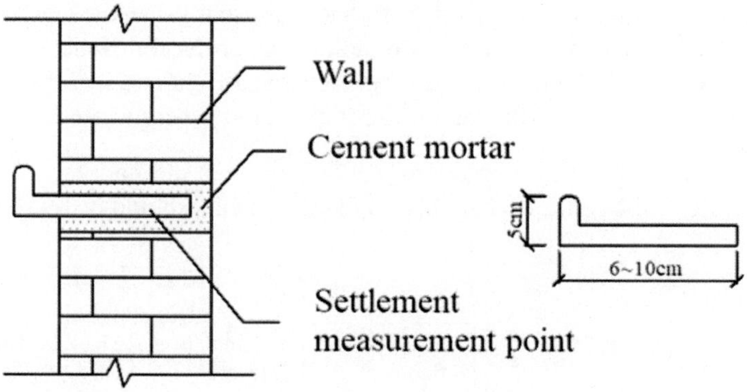

Fig. 2.4 A schematic diagram of settlement measurement points of buildings

Table 2.3 Main technical requirements for precision leveling

Elevation difference per kilometer Median error (mm)		level grade	Level ruler	Observation times	Bad round trip, attachment, or Loop closure difference (mm)
Accidental mean error	Complete mean error	DS1	Indium steel rule	Round—trip measurement once each	$4\sqrt{L}$
±1	±2				

Note L is the route length (in km) of the round trip section and loop line;

Fig. 2.5 Measuring instrument

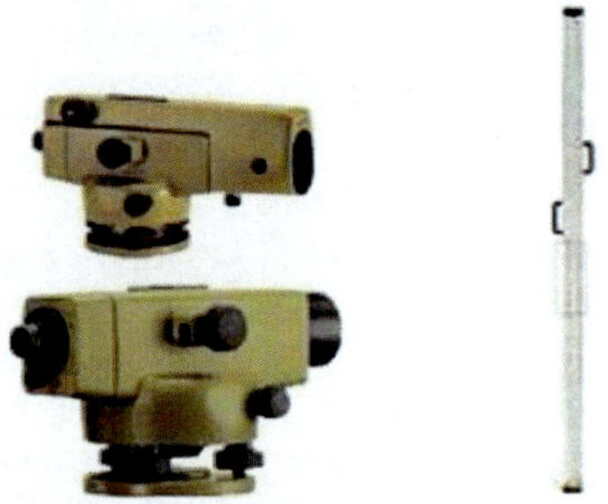

(a) WILDNA2 automatic leveling level (b) Indium steel rule

iii. Observation method: "back-front-front-back" is used for odd-numbered stations, and "front-back-back-front" is used for even-numbered stations. For odd-numbered stations, "front-back-back-front" is used, while for even-numbered stations, "back-front-front-back" is used. When the measurement to the back of the conversion, the two indium steel ruler for exchange;
iv. The line-of-sight length, the difference between front and rear stadia, and the line-of-sight height of the measuring station are shown in Table 2.4.

v. See Table 2.5 for the observation limit difference at the station.
vi. The height difference exceeding two measurements should be retaken.

Table 2.4 Field observation requirements

Rod type	Length of the line of sight		Stadia difference between front and rear	Accumulative difference between front and rear stadia	The line of sight height	
	Instrument rating	Stadia			Line of sight over 20 m	Line of sight below 20 m
Indium steel rule	DS1	50 m or less	1.0 m or less	3.0 m or less	0.5 m	0.3 m

Table 2.5 Measurement station observation limit difference

Kiev division reading difference	The Kiev division the difference between the elevation difference	Mean value of upper and lower wire reading and medium wire The difference between the readings	Detect the difference in interval point height
The 0.5 mm	The 0.7 mm	The 3.0 mm	The 1.0 mm

3) After the completion of the observation of the settlement reference network, the measurement records shall be strictly checked, and the difference between each leveling closing ring shall be controlled as far as possible. The adjustment calculation of the internal industry shall be carried out only after all parameters are recognized as qualified. In general, adjustment software is used to carry out strict adjustment calculation according to indirect adjustment method, and the unit of elevation result can be as high as 0.1 mm.

(2) Settlement measurement at monitoring points

According to the requirements of national second-class leveling standards, each settlement observation is conducted by measuring a corresponding second-class leveling closed route between working base points, and the displacement value of each monitoring point is observed by working base points along the route. The initial value of the elevation value of each monitoring point should be measured twice in the early stage of the project to get the average value. The settlement value of a monitoring point this time is the difference of the elevation value minus the previous elevation value, and the accumulated settlement value is the difference of the elevation value minus the initial elevation value.

(3) Monitoring frequency

Due to the dynamic nature of shield tunneling, monitoring also requires synchronous dynamic tracking. The focus should be on monitoring all monitoring points within the length range of 20 in the first stage and range of 30 m in the last stages of shield tunneling.

The monitoring period and frequency during shield tunneling of key buildings are shown in Table 2.6.

Table 2.6 Monitoring frequency of key buildings

The construction phase	Shield advance		
Construction area (distance from building)	10–30 m	Within 10 m	30 m away
Monitoring frequency	1–2 times/1 day	2–4 times/day	Gradually stop test

Table 2.7 Frequency of construction monitoring

	Shield drive promotes all pipelines, buildings and surface monitoring points within 20 m in the front and 30 m in the rear of the construction section	All pipelines, buildings and surface monitoring points within 10 m before and 10 m after the shield drive construction section	30 m external pipelines, buildings and surface monitoring points behind shield drive construction section
Shield advance	At least 2 initial values		
The shield is advancing	1 time / 1 day	2 times per day	
As the shield advances			Once /3 days (0–10 days) Once /7 days (10–30 days) Phasing out (30 days)

The monitoring period and frequency of interval shield tunneling are shown in Table 2.7.

(4) Alarm value

According to relevant codes and design requirements, the cumulative alarm value of ground and building settlement is set as (+10 mm ~ −30 mm), and the alarm value of single uplift sinking is ±3 mm; The tilt alarm value of the building is $\delta/l < 3/1000$ (δ is the differential settlement value, l is the length of the building). When the monitoring point reaches the alarm value, it will immediately give an alarm, analyze the reasons and take corresponding technical measures.

2.3.2 *Arrangement of Measuring Points*

(1) Monitoring of the ground and buildings around the Genshanmen station

Shield crossing the first automobile company, the modern home, just out of the hole for shield, some construction parameters did not adjust to the best, plus shield posture, also is not very good, the surface and buildings are greatly influenced by the shield tunneling, so need to monitor the first automobile company, modern home and the subsidence of the ground along, measuring points arrangement is shown in Fig. 2.6. The First Automobile Company is a short pile foundation building, and the modern

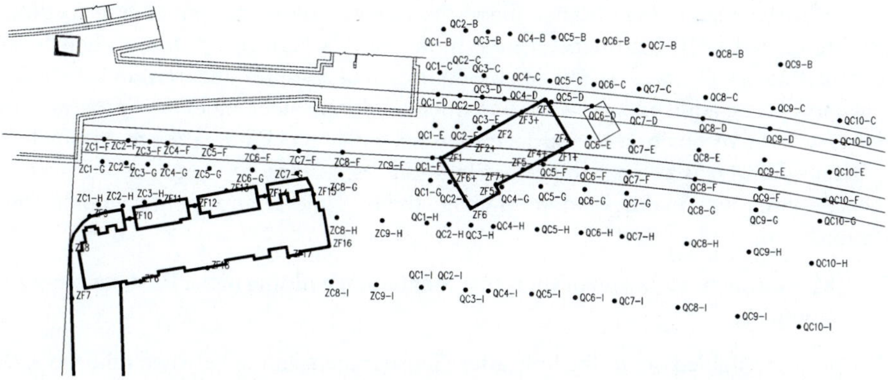

Fig. 2.6 Monitoring points layout of the ground and buildings around the Station

home residential building is a pile foundation building. The specific layout of the measuring points is as follows:

1) Layout of surface measuring points

Transverse layout: infill monitoring is carried out within 90 m from shield tunneling Wells (exit section and entry section). Fig. 2.7 shows the specific layout of transverse settlement section as follows: one section is arranged every 5 m within 0–15 m; A section is arranged every 10 m from 15 to 45 m; 45 to 90 m, each 15 m to arrange a section; The rest of the section is arranged every 50 m. There are altogether 32 sections (No. QC1~QC32) in the interval from Genshanmen station to Zhanongkou station. Section monitoring range (section length) is 30 m outwardly from the center of each tunnel. A total of 9 measuring points are arranged on each section, and the remaining 6 measuring points are 3 on both sides. The distance from the measuring points on the center line of the two tunnels to the is 5 m, 10 m and 15 m, respectively (A total of 30 m). The measuring points of each section group are numbered from A to I.

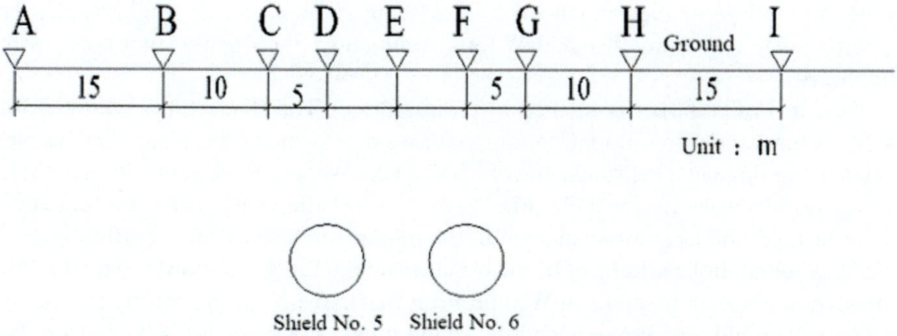

Fig. 2.7 A schematic diagram of measuring points in transverse profile of surface settlement

The surface settlement points above the tunnel axis adopt embedding method, drilling drilled 5–10 cm diameter hole in the earth's surface, break the shell layer after nailing 60–80 cm (according to the condition of complex underground obstacles and appropriate adjustments) with sand filled after long steel, other area general with nut type reserve position the surface of steel nails, special area (such as pipeline concentration areas or near important building) according to the above the axis of the stationing way to deepen don't damage the pipeline arrangement of monitoring stations.

2) Measuring point layout of short pile foundation building of the First Automobile Company

The project shielded out of the hole after Genshanmen station between mileage K18 + 40 ~ K18 + 115 from the man under coach group office building, the office building is 6 layer brick structure, foundation for thin-wall Φ 500 mm prestressed pipe pile, pile length 8.5 m, the pile bottom elevation of −3.5 (yellow sea elevation). The top elevation of No.5 and No.6 shield is −5.385, which is about 1.9 m away from the bottom of pile foundation and 14.5 m away from the center of the tunnel. The shield tunneling through the soil is $④_3$ Layer silt silt clay contains silt clay and $⑥_1$ Layer silt silty clay. 13 monitoring points are arranged on the building, with an average distance of about 7 m and a minimum distance of about 2 m, as shown in Fig. 2.8.

3) Measuring point layout of modern home long-pile foundation buildings

The No. 6 shield tunnel of the section passes through the modern home of Yefeng Real Estate after it is excavated from Genshanmen Station, building foundation for Φ 600 mm and Φ 700 mm of bored piles, pile length 39.8 m, about 6 m from tunnel line at least, Sect. 6 shield at the top of the buried depth about 10 m, through the soil for $③_6$ and $③_7$ Layer silty sand and sandy silty soil. A total of 13 monitoring points are arranged on the building, among which 8 points are located on the main building and 5 points are located on the upper skirt building, as shown in Fig. 2.9.

(2) Measuring point layout of shallow foundation buildings near Hangzhou Wenhui Fruit Market

The shallow foundation buildings in the area affected by shield tunneling are more susceptible to the influence of shield construction and are highly sensitive to settlement. Therefore, the shallow foundation above the shield is monitored with emphasis.

Wenhui Fruit Market is located in the middle of The Station, the west side of which is the Railway of Freight station, the east side of which is the residential district "Tiancheng Jiayuan", the south side of which is the Wenhui Bridge, in addition, there are many old houses around it. See Fig. 2.10 for the relationship between the tunneling route of no.5 shield and the surrounding structures of Metro Line 1. Buildings A (shallow foundation building of Railway Ganshanmen Freight station) and B (shallow foundation building of shops in Wenhui Fruit Market) are approximately parallel to the axis of shield tunnel No. 5, and the shield tunnel passes under it. Therefore, the

(a) Photos of the office building of the coach team

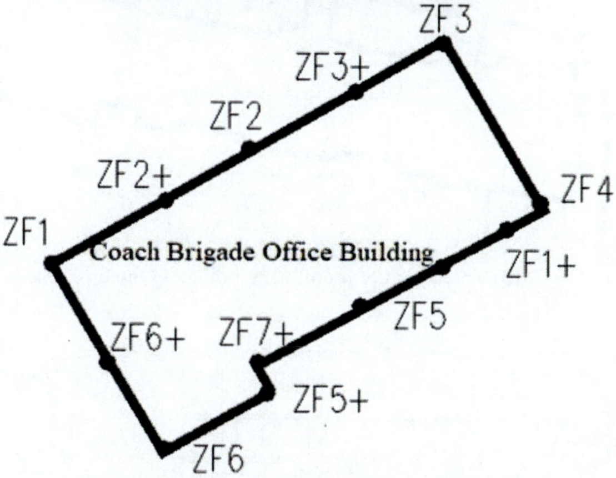

(b) The layout of monitoring points in the building

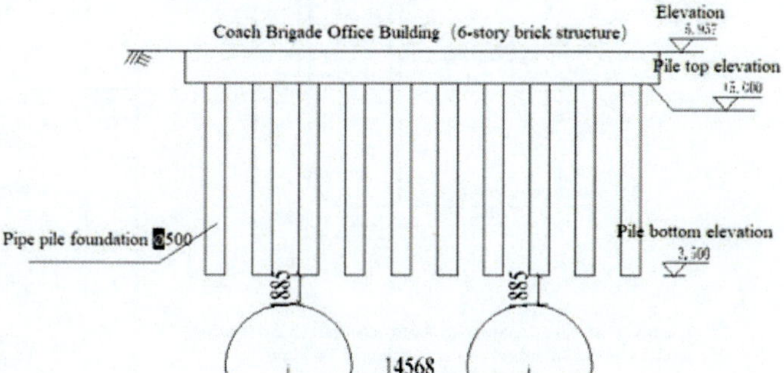

(c) The location relationship between the building and the subway tunnel

Fig. 2.8 Short pile foundation building of First Automobile Company

(a) Photographs of modern homes

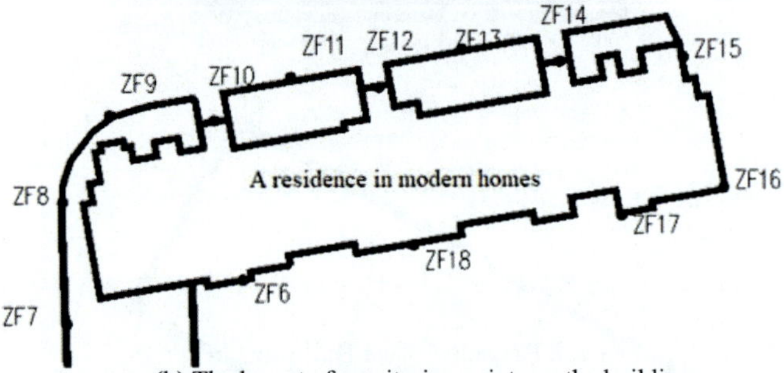

(b) The layout of monitoring points on the building

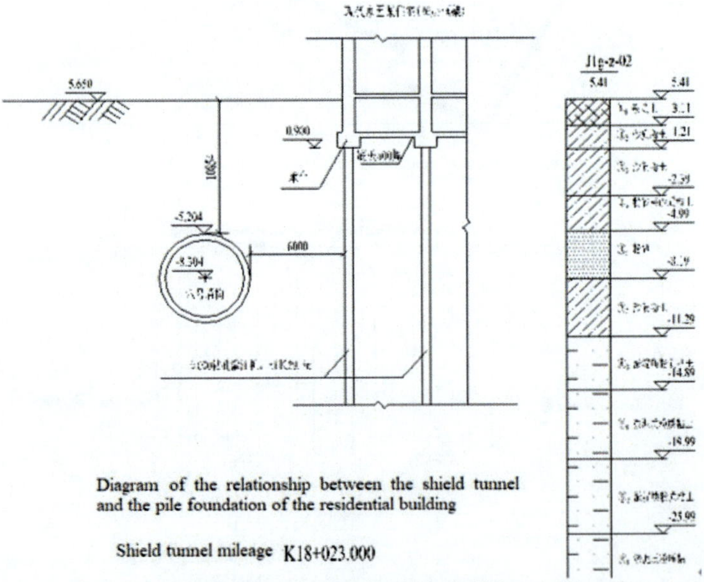

(c) The relationship between the location of the building and the subway tunnel

Fig. 2.9 Modern home long pile foundation building

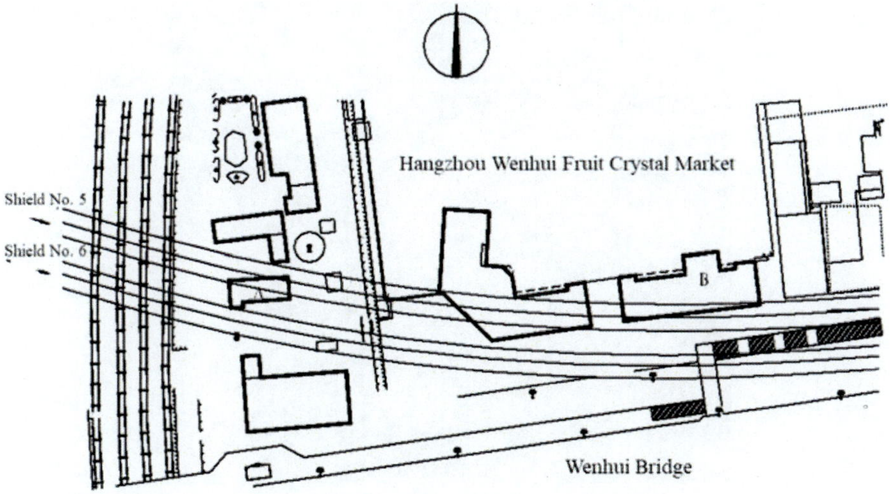

Fig. 2.10 Location relationship between tunnel and surrounding buildings

monitoring data of the shield tunnel is analyzed with emphasis. Building A is a two-storey brick structure with a strip foundation. The buried depth of the foundation is about 0.5 m. Along the axis of the tunnel, the length is about 13 m and the width is about 6 m, 4 measuring points are arranged, numbered as TF24-1~24-4 (See Fig. 2.11). Building B is a two-storey brick and concrete structure with a strip foundation. The buried depth of the foundation is about 0.5 m. Along the axis of the tunnel, the length is about 31 m and the width is about 10 m, 6 measuring points are arranged, numbered as QF08~13 (See Fig. 2.12).

(a) Photographs of the building A

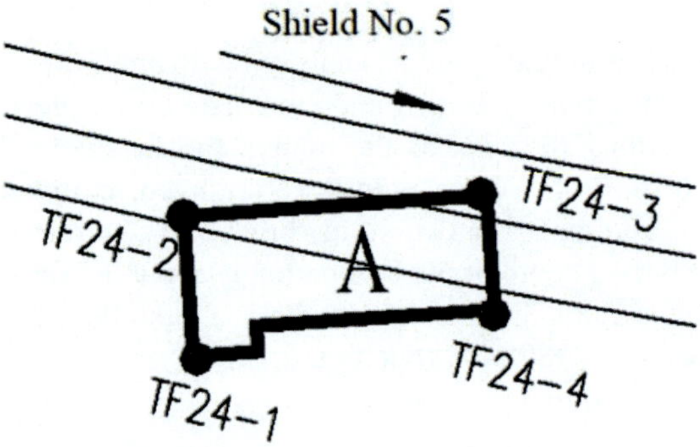

(b) The layout of monitoring points in Building A

Fig. 2.11 Shallow foundation building of Railway Freight station at Genshan Gate

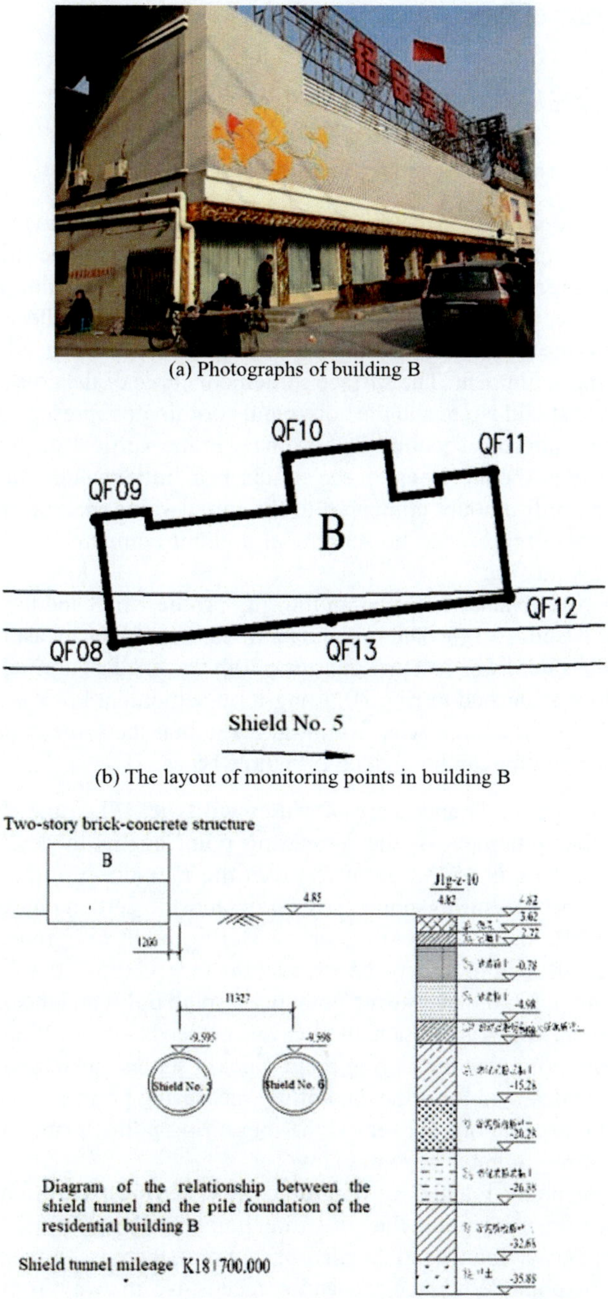

(a) Photographs of building B

(b) The layout of monitoring points in building B

(c) The location relationship between the house and the subway tunnel

Fig. 2.12 Shallow foundation buildings of shops in Wenhui Fruit Market

2.4 Analysis of Measured Results

2.4.1 Analysis of Measured Land Settlement Curve

(1) Settlement changes of surface measurement points

The horizontal coordinate is the distance between the shield excavation surface and the transverse monitoring section. A positive value means that the shield excavation surface has not reached the measuring point, and a negative value means that the excavation surface has left the measuring point. The ordinate is the settlement value of the surface at the measuring point, with positive value indicating uplift and negative value indicating settlement. The surface settlement curve of the cross section during the tunneling of shield No.5, and the horizontal coordinate represents the monitoring point on the monitoring section; The ordinate is the vertical displacement value, which is defined as the increment of the comparison initial value. The positive value represents the uplift amount compared to the initial value measurement point, and the negative value represents the settlement amount compared to the initial value measurement point.

The horizontal distance between monitoring profile No. 3 and the shield starting working well is about 15 m. The initial data in the figure was measured on June 20, 2009, when the shield was just jacking. June 30th was the last normal measurement, and the shield was pushed to ring 30. Long-term settlement book is not considered for the time being. It can be seen from Fig. 2.13a that the settlement change curve of the measuring point can be divided into three types:

1) Measuring points D and E are of settlement type. When the shield tunneling begins, the settlement of the measuring point begins to develop. When the excavation face is 15–3.5 m away from the measuring point, the settlement speed of the measuring point E is faster and the settlement amount reaches 7 mm, while the settlement of the measuring point D is relatively slow and small, about 2 mm. When the excavation face is 3.5~−5 m apart from the measuring point, the measuring point has a small uplift and the measuring point E is relatively obvious. When the excavation face is −5 ~ −12 m away from the measuring point, the measuring point quickly settles again, and the measuring point E settles by 3 mm (the data of the measuring point D is not considered). When the excavation face leaves the measuring point 12 m, the settlement of the measuring point E becomes slow.
2) Measuring points B and C are of uplift type. When the excavation face is 15–7 m from the measuring point, the measuring point is slightly uplifted; Within 7–4 m from the excavation face to the measuring point, there is 1 mm settlement at the measuring point; When the excavation face is 4–5 m away from the measuring point, the measuring point uplift is 2 mm. After that, the deformation of the measuring point is relatively stable. Until the excavation face is 13 m away from the measuring point, the uplift of the measuring point increases, and the uplift of the measuring point D is relatively obvious.

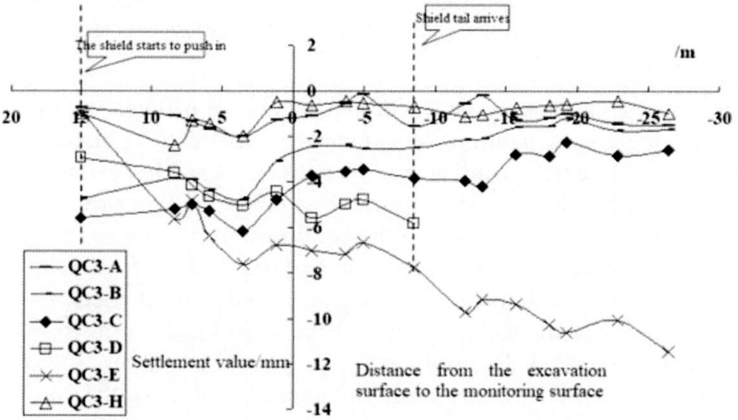

(a) Settlement process of QC3 profile monitoring point

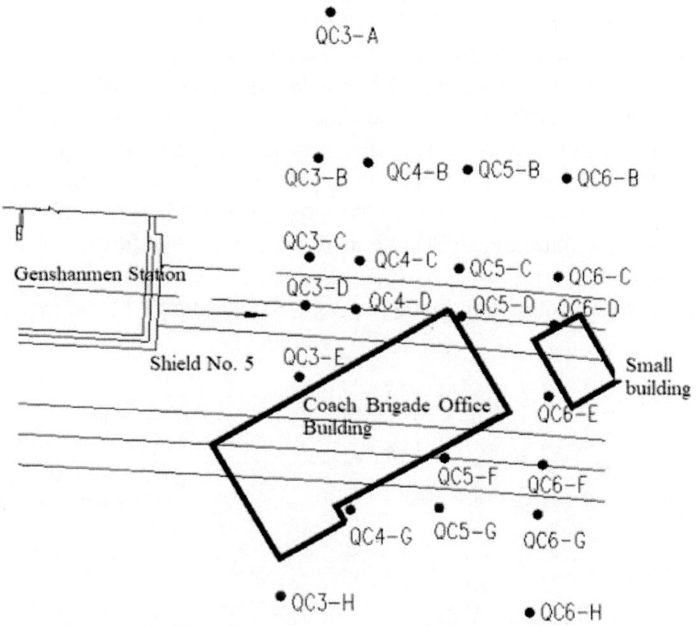

(b) The location relationship between the coach office building and the tunnel

Fig. 2.13 Surface subsidence

3) Measuring points A and H are stable. The deformation of the measuring point is in dynamic equilibrium, so it can be considered that it is almost not affected by shield tunneling.

In general, the attitude of the shield machine is not particularly good when the shield tunneling is carried out, so deviation correction control is needed, which will cause a certain degree of disturbance to the surrounding soil, specifically, the settlement trend of each measuring point is not consistent, which will have a certain impact on the settlement and deformation of the building above and its pile foundation. In the construction process, the propulsion axis of the shield can be controlled by the combination of automatic measurement system and manual check to ensure that the tunneling axis of the shield will not deviate too much, so as to minimize the disturbance of the surrounding soil.

The horizontal distance between no.5 monitoring profile and the starting working well of shield is about 30 m. The measurement points of this profile were measured on June 25, 2009, for the first time, when the shield was pushed to ring 5. On July 8th, the last normal measurement, the shield was pushed to ring 83. It can be seen from Fig. 2.14 that before the shield arrives at the measuring point, the measuring points C and F are slightly uplifted, while the measuring point D is slightly subsided. When the shield tunneling through the measuring points, the measuring points show uplift. When the shield tail leaves the measuring point (the shield tail reaches the measuring point), all the measuring points will settle. In particular, the settlement speed of measuring point D is fast, which will rebound after reaching the maximum settlement of 6 mm; the settlement of measuring point F is slow, which will rebound after reaching the maximum settlement of 3.5 mm; the deformation of measuring point C is relatively stable. Due to the increase of the distance from the entrance, the trend of surface settlement maintains a good consistency. Considering the further

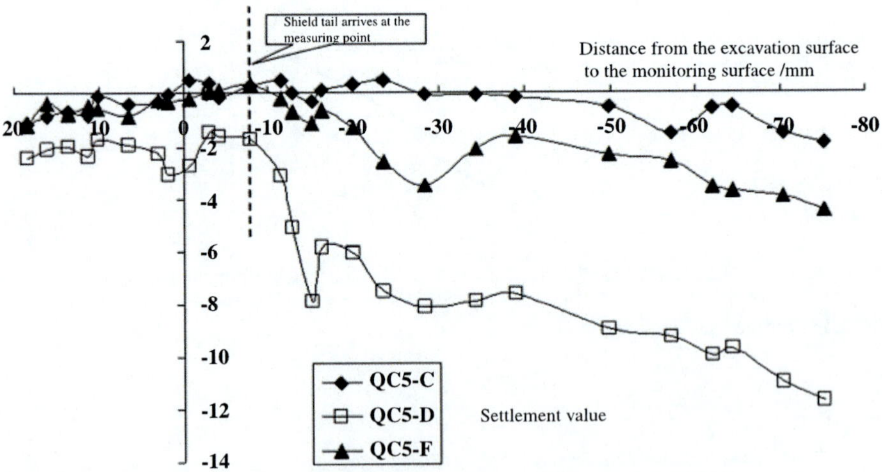

Fig. 2.14 Settlement process of some monitoring points in QC5 profile

development of the settlement and the proximity to the pile foundation of the building (1.9 m), the excavation rate is well set, which is generally controlled within 99%. At the same time, the grouting pressure of shield tail is specially controlled, which is generally less than 0.4 mpa to prevent the soil disturbance degree from being large. Considering the loss of slurry, the grouting rate is generally controlled between 150 and 200%. Due to the existence of short pile buildings, the surface settlement deformation is large, and secondary grouting is carried out at the same time during the actual construction, and the grouting pressure is generally less than 0.4mpa.

The horizontal distance between monitoring profile No. 6 and the starting working well of shield is about 40 m. The measuring points of this profile were first measured on June 25, 2009, when the shield was pushed to ring 5. On July 8th, the last normal measurement, the shield was pushed to ring 83. Figure 2.15a shows the settlement curve of some measuring points on Sect. 6 when Shield 5 is pushed forward. When the excavation surface is not up to the measuring points and the distance is greater than 12 m, the measuring points are slightly uplifted but not significantly deformed, and the impact of shield tunneling is very small. When the excavation face is 12 m away from the measuring point, the measuring point will be uplifted, and the uplift will increase gradually with the shield excavation face approaching the measuring point. When the excavation face reaches the measuring point, the uplift rate of the measuring point suddenly increases. When the excavation face is 4 m away from the measuring point, the uplift suddenly stops. But then the second grouting allows the rapid development of the uplift again. When the shield tail separates from the measuring point (when the shield tail reaches the measuring point), the uplift value of the measuring point immediately reaches the peak value. The total uplift amount of the measuring point in the previous process is about 4 mm, and then the measuring point rapidly sinks. Then, through the secondary grouting construction, the measurement point is rapidly uplifted again. When the excavation face is about 30 m away from the measurement point, the uplift quantity reaches the peak again. Finally, as the shield tunneling machine is far away, after the excavation face is 48 m away from the measuring point, the measuring point begins to settle, which is mainly caused by the drainage and consolidation of soil.

It can be known from the above: (1) When the shield tunneling, due to the unbalance of the excavated soil, the knife disk will squeeze the soil in front and make the front surface uplift; In the process of the shield crossing the measuring point, the uplift caused by the extrusion of the fuselage against the soil above is very obvious. Back-wall grouting and secondary grouting can fill the gap caused by shield excavation, but if the amount of grouting is too much, the surrounding soil will also be squeezed, resulting in surface uplift; If the grouting amount and grouting pressure are appropriate, the surface settlement can be effectively controlled; At the same time of grouting, there is settlement caused by soil disturbance and stress release; The deformation of the uplifted measuring point has a great relationship with the unearthed quantity, grouting quantity and grouting rate.

Figure 2.15b shows the transverse surface settlement change curve of monitoring section No. 6. It can be seen that the surface deformation during shield tunneling is mainly manifested as uplift. As can be seen from the figure, when the shield is

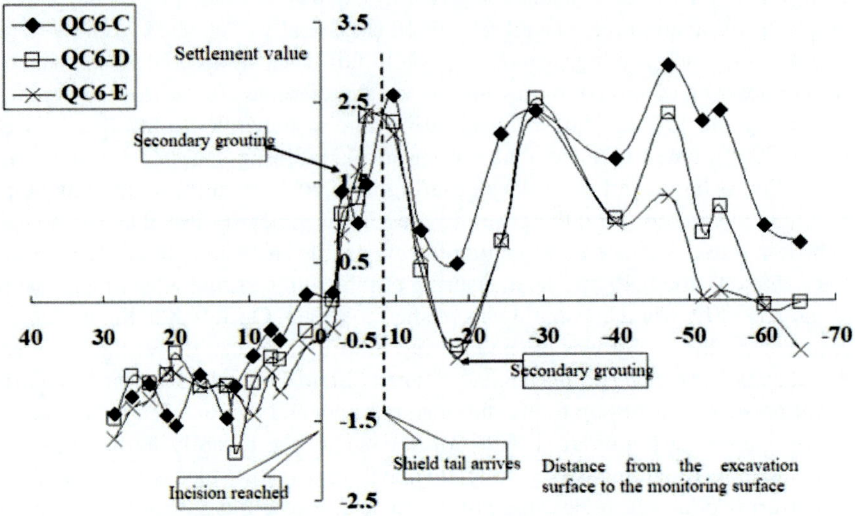

(a) Settlement process of some measuring points in SECTION QC6

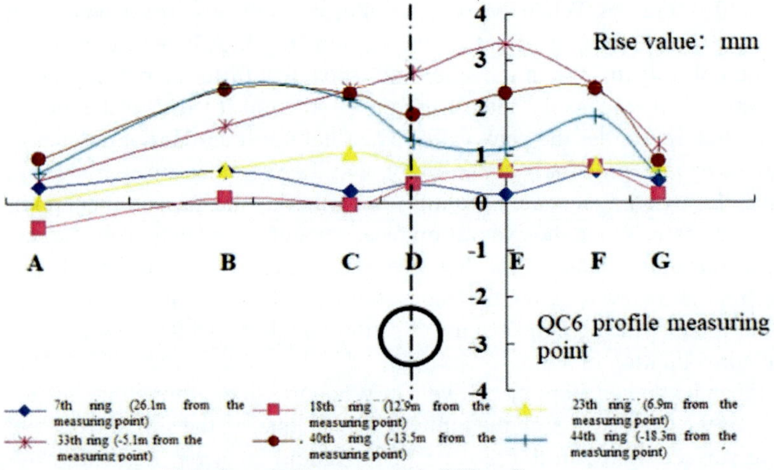

(b) Surface subsidence changes in QC6 profile

Fig. 2.15 Surface settlement curve

pushed to ring 7 (when the excavation face is far from the measuring point), the uplift amount of the transverse settlement curve is relatively uniform and small, which can be considered as measurement error, but other possibilities cannot be ruled out. When its push to 18 ring tunnel is directly above and have a slight bulge on the right side of the points, along with the advancement of shield, the uplift of the ground gradually increase, but in the process of shield near point D, E, the amount of uplift are greater than the adjacent point however, until the shield through the process of measuring point, D, E, uplift to behave slightly larger than the adjacent point; When the shield tail is separated from the measuring point, the uplift amount decreases, and the settlement of measuring points D and E is obviously larger than that of adjacent measuring points, and the transverse surface settlement curve is concave between D and E.

According to the book, when the shield is near the stage, due to the unbalance of excavation, the soil in front will be slightly disturbed and produce a small uplift, but the building in front will weaken the uplift. When the shield tunneling passes through the measuring point, the soil around the shield tunneling machine is greatly disturbed, and the buildings directly above the shield obstruct the movement of the soil below, which makes the soil around the building more easily uplifted, as shown by the larger amount of uplift of the measuring points on both sides of the building. When the shield is leaving the stage, the soil sandwiched between the building and the shield will release the stress rapidly, resulting in a large settlement. The subsidence of the building will drive the surrounding soil to settle, which is manifested as a faster decrease in the uplift amount of the measuring point.

Monitoring profile No. 7 was about 50 m from the horizontal distance of the shield's initial working well. The measuring points of this profile were first measured on June 27, 2009, when the shield was pushed to ring 11. On July 10, the last normal measurement, the shield was pushed to ring 95. As shown in Fig. 2.16a, when the excavation face is about 20 m away from the measuring point, the measuring point presents an uplift trend. As the shield is close to the measuring point, the uplift gradually increases. When the shield tunneling through the measuring point, the measuring point uplift is faster. When the shield tail is separated from the measuring point, the uplift amount of the measuring point increases rapidly due to the action of grouting behind the wall. The uplift value of the measuring point D above the tunnel axis is about 4 mm, slightly larger than the adjacent measuring points C and E. It can be seen that grouting has a great influence on the surface deformation. Figure 2.16b shows the surface settlement change curve of SECTION QC7. It can be seen that, under the influence of shield tunneling, the amount of uplift of soil directly above the shield is larger than that on both sides of the tunnel.

The horizontal distance between monitoring profile No. 8 and the starting working well of shield is about 65 m. The measuring points of this profile were first measured on June 30, 2009, when the shield was pushed to ring 27. On July 13, the last normal measurement, the shield was pushed to ring 115. As shown in Fig. 2.17, when the surface at the measuring point is affected by shield tunneling, the uplift is dominant, and the maximum uplift amount is about 4 mm. The excavation surface rebounded after leaving the measuring point 40 m, and then the surface continued to settle.

The horizontal distance between monitoring profile No. 9 and the shield starting working well is about 80 m. On July 3, 2009, the measuring points of this profile were measured for the first time, when the shield was pushed to ring 44. On July 13, the last normal measurement, the shield was pushed to ring 115. Figure 2.18 shows the settlement change curve of some measuring points in section QC9. It can be seen that during the approach and crossing of the shield, the measuring points have an

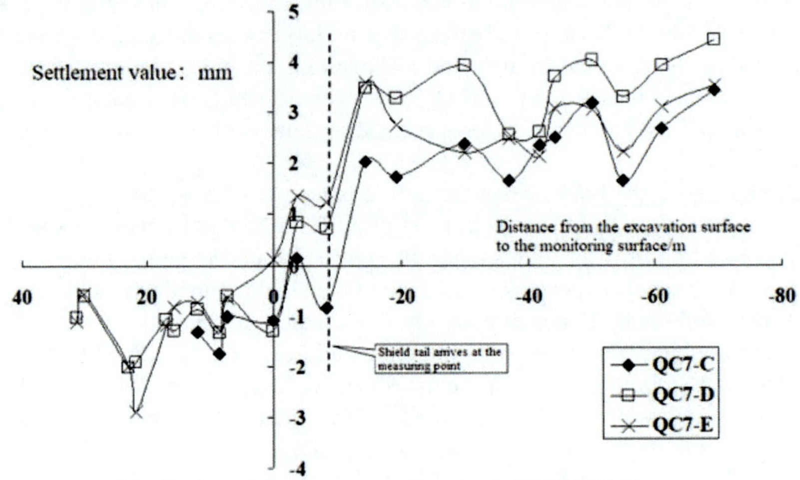

(a) Settlement process of some measuring points in SECTION QC7

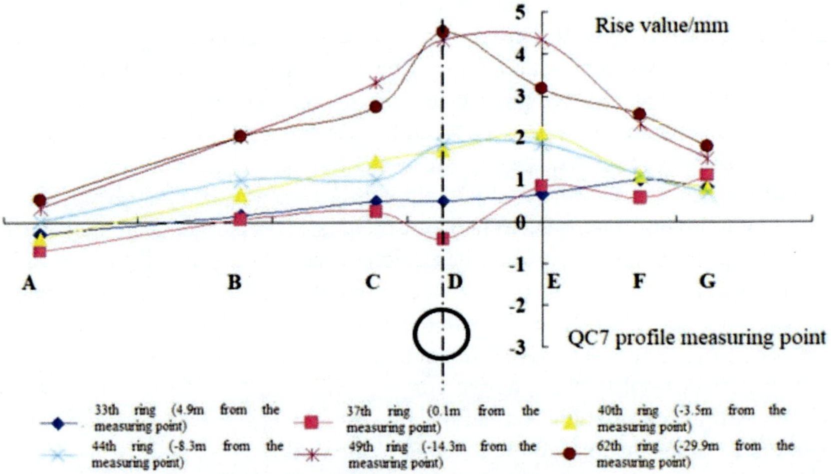

(b) Surface subsidence changes in QC7 profile

Fig. 2.16 Surface settlement curve

uplift of about 1 mm. When the shield tail reaches the measuring point (the shield tail leaves the measuring point), the ground begins to settle. When the excavation face is about 25–45 m away from the measuring point, the settlement of the measuring point bounces slightly.

The horizontal distance between the monitoring profile No. 10 and the starting working well of shield is about 95 m. The measuring points of the profile were first measured on July 7, 2009, when the shield was pushed to ring 72. On July 13, the last normal measurement, the shield was pushed to ring 115. Figure 2.19

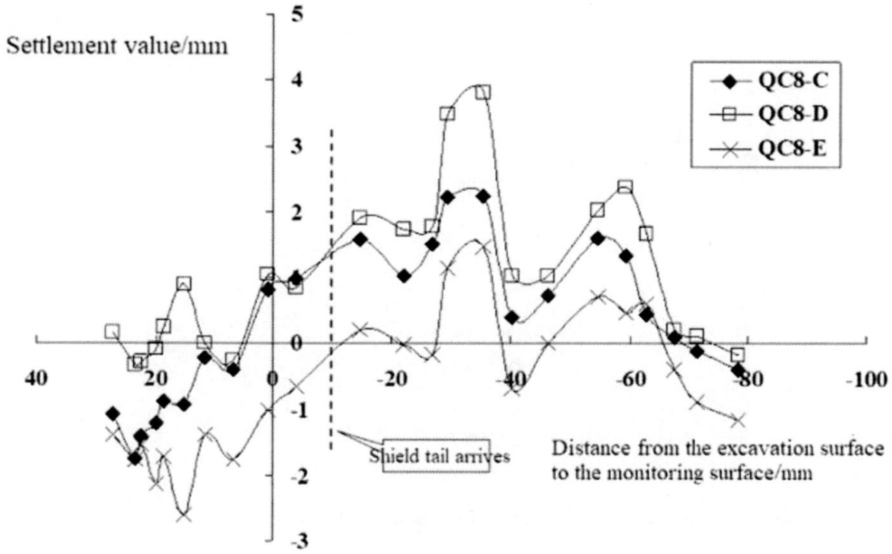

Fig. 2.17 Settlement process of some measuring points in QC8 profile

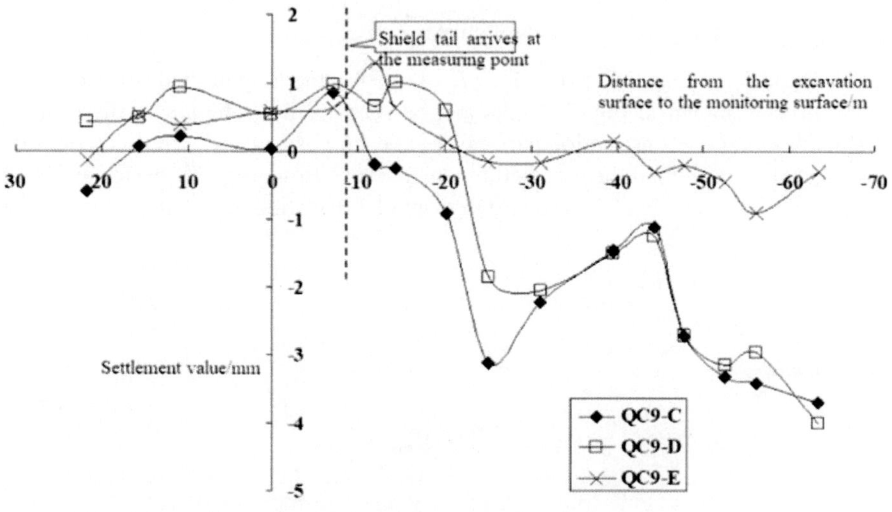

Fig. 2.18 Settlement process of some measuring points in QC9 profile

shows the settlement change curve of some measuring points in QC10 profile. It can be seen that the surface of the measuring points is slightly uplifted when the shield tunneling crosses the measuring points. When the shield tail block leaves the measuring point, the measuring points C and D above the tunnel begin to settle. The

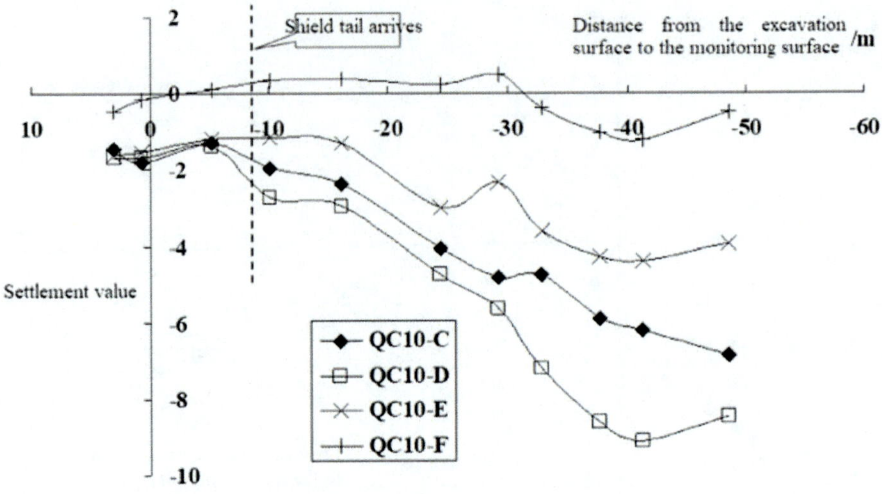

Fig. 2.19 Settlement process of some measuring points in QC10 profile

settlement rebounded slightly when the excavation face was 25–30 m away from the measuring point.

(2) Longitudinal surface settlement curve

When shield 5 is advanced to ring 30, 37, 44 and 49, the longitudinal surface settlement curve is shown in Fig. 2.20. As can be seen from the figure, in the process of shield tunneling, construction parameters at each position are different, resulting in different longitudinal surface settlement curves. However, the basic law is as follows: the soil outside of 7 m square in front of excavation is less affected, and the

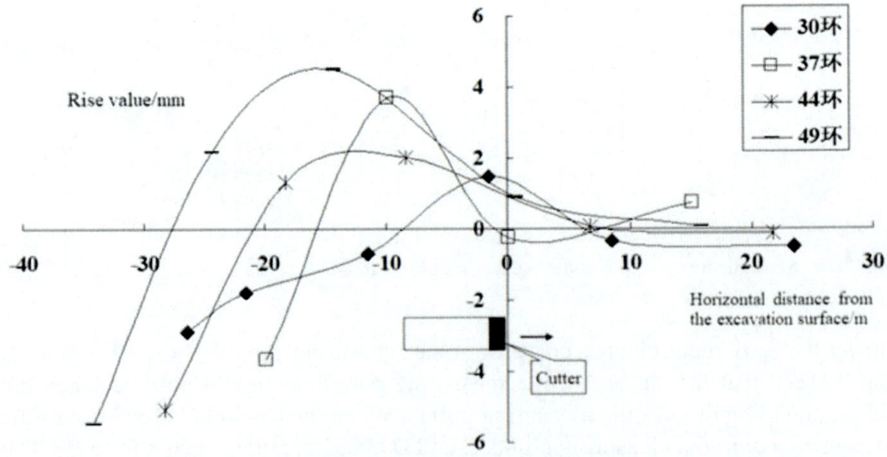

Fig. 2.20 Longitudinal surface settlement curve

soil within 7 m has a small amount of uplift; The square soil on the shield machine is still uplifted; The uplift of the ground behind the tail of the shield will generally reach its maximum. After a distance from the tail of the shield, the deformation of the ground will eventually be settlement. Among them, the excessive uplift at the shield tail of ring 37, 44 and 49 is mainly caused by secondary grouting, while the uplift at the excavation face of ring 30 is mainly caused by unbalance of excavation.

There are many reasons for surface uplift, Shaoming [1] summarized the following five situations: (1) The undercut state of the excavation surface, that is, the volume of the cut soil is greater than the volume of the discharged soil, resulting in negative formation loss, which causes the surface uplift in front of the shield machine; (2) The supporting pressure of the excavation face is too large. When the set value of the earth pressure of the excavation face is greater than a certain limit of the static earth pressure of the excavation face or reaches the passive earth pressure, the square earth body in front of the excavation will move or be destroyed in the direction of about 45° above the front of the shield, leading to a large number of uplift or slurry and mud in front of the shield machine. (3) The synchronous grouting pressure is too large and the grouting amount is too much, which causes the surface uplift of the shield tail; (4) When there is an obstacle at the incision of the shield, the shield will push it to move forward with the shield, leading to the ground uplift in front; (5) When shield tunneling through clay with high viscosity, especially when double shield tunneling is advanced, it is easy to drag the clay on its back or retain clay with high cohesive force, resulting in backsoil effect. It is considered in this book that the uplift of excavation surface is mainly caused by unbalance of excavation, which can be caused by underexcavation of excavation surface, excessive support pressure and the comprehensive effect of backsoil.

2.4.2 Settlement Curve Analysis of Different Foundation Buildings Measured

The soil deformation above the tunnel is related to the deformation of the adjacent buildings, and the existence of the adjacent buildings restricts the movement of the soil, which is related to the dead weight and foundation stiffness of the buildings. For example, the overall stiffness of pile-raft foundation is better than that of strip foundation and independent foundation, so it can better resist various forms of formation deformation such as settlement, differential settlement and horizontal displacement. Therefore, it is of great significance to study the deformation law of buildings with different foundation forms in the process of shield tunneling.

(1) First Automobile Company (Short pile foundation)

Shield 5 passes under the north corner of the office building of coach Brigade of the First Automobile Company (see Fig. 2.21). The settlement measurement results of each measuring point are shown in Fig. 2.22. It can be seen from the figure that the settlement amount of measuring points ZF1, ZF2+, ZF3+, ZF2, ZF6+ and ZF3

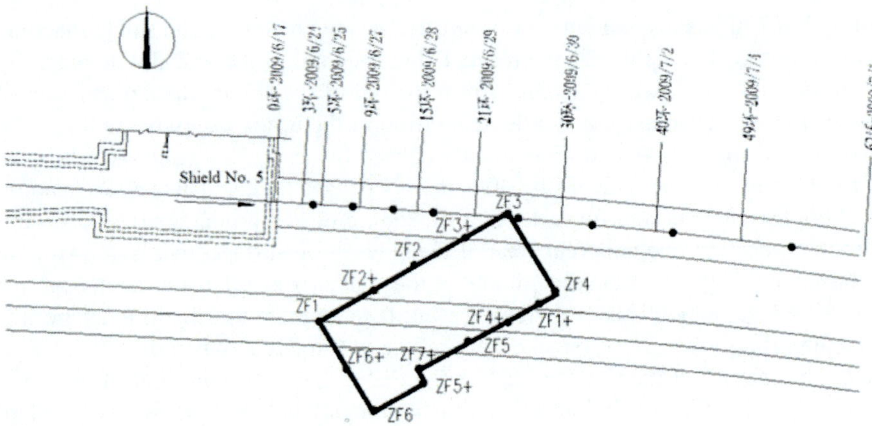

Fig. 2.21 Changes in the tunneling position of shield

is larger than that of other measuring points, and the relation of their maximum settlement amount is: ZF1>ZF2+>ZF3+>ZF2>ZF6+>ZF3; Measuring points ZF1, ZF2+, ZF2 and ZF6+ are close to the starting well and affected by shield tunneling relatively early. On June 21, 2009, their subsidence began to develop rapidly. Among them, measuring points ZF1, ZF2+ and ZF6+ reached large values and rebounded somewhat on June 23. The final settlement of the east corner of the building is 0.5 mm, the south corner is 4 mm, the west corner is 18 mm, and the north corner is 8 mm. It can be seen that the whole of the building slopes along the east–west direction, which is not only influenced by the unbalance and grouting during the excavation of the shield tunneling, but also closely related to the location of the building and tunnel.

It can be seen from Fig. 2.23 that the office building of the coach brigade tilted during the shield crossing. The reasons are as follows: The settlement of ZF1 and ZF2+ occurred relatively early, while the settlement of ZF3 has not yet started or just started; (2) Due to the unbalance of excavated surface of shield tunneling machine, the basement of buildings in the adjacent tunnel was subjected to high soil pressure, which made the settlement smaller or produced uplift; The building has a large stiffness, which makes the settlement of the building more uniform and the whole inclined to one side.

Comparing the two pictures, it can be found that the settlement on the north side of the building is larger than that on the south side, and the whole building slopes along the east–west direction. In addition, the law of settlement and tilt of the building is as follows: the building is first affected by shield construction, resulting in a large settlement, and the early settlement develops at a faster rate. When the settlement reaches 13 mm or so, the settlement becomes slow. At this time, the settlement of the eastern corner just begins to develop and develops at a faster speed. When the settlement reaches 7 mm, it slows down somewhat. The settlement in the middle of the building is generally more uniform.

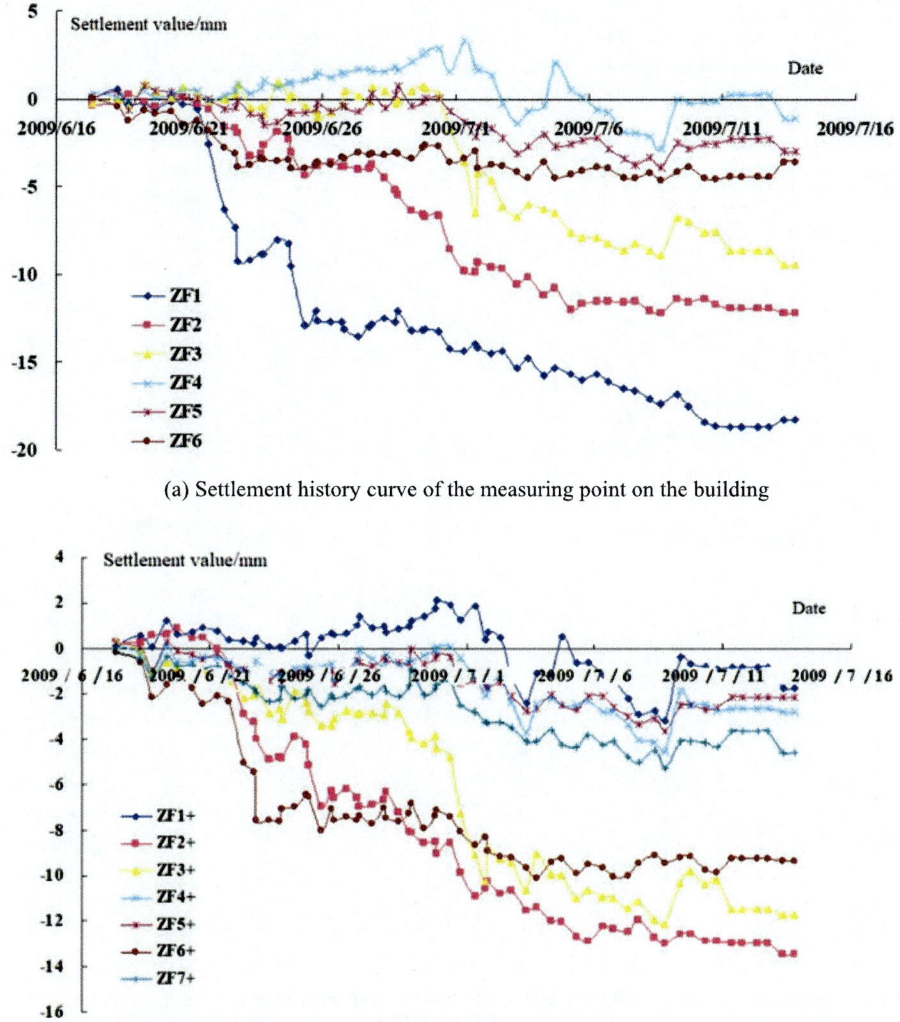

(a) Settlement history curve of the measuring point on the building

(b) A settlement history curve for the addition of monitoring points on the building

Fig. 2.22 The settlement of the building does not change during the tunneling of shield 5

As shown in Figs. 2.22 and 2.23, the measured results show that the structure deformation caused by crossing the short pile foundation building below is relatively large, so the excavated volume of shield and the synchronous grouting volume of shield tail should be strictly controlled, and secondary grouting can be carried out if necessary. The excavation rate should not be less than 98%, and the overbreak should not be formed. The grouting rate should be controlled within the range of 160–200%, the pressure of the earth bunker on the tunneling surface of the shield should be controlled within the range of 0.2 Mpa, and the grouting pressure should

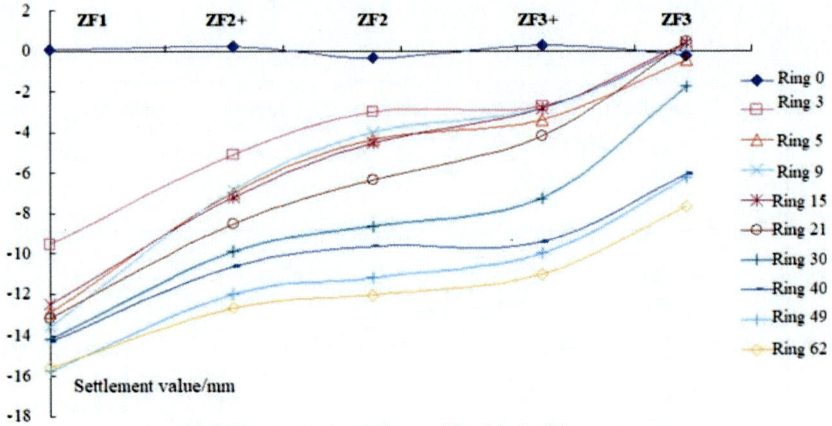

(a) Settlement on the northwest side of the building

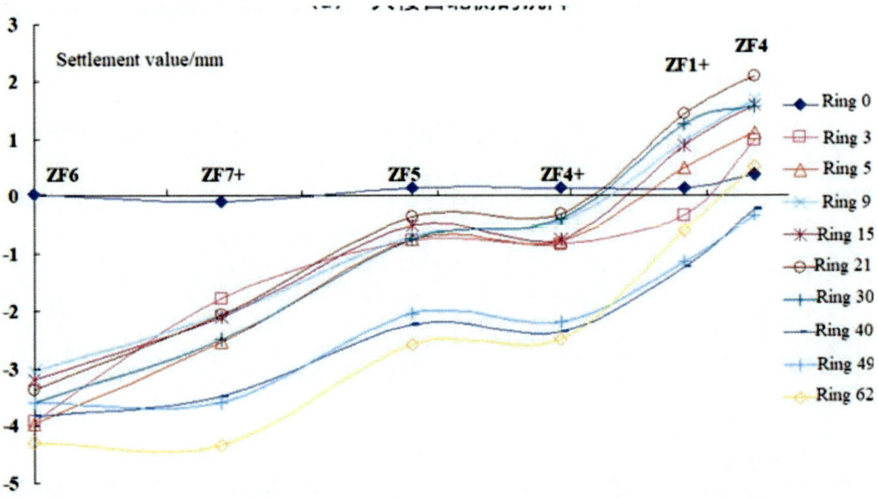

(b) Settlement on the south-east side of the building

Fig. 2.23 Deformation of the building

be controlled within the range of 0.3–0.5 Mpa. When the shield tunneling machine is crossing, the excavation surface of the shield should have a small uplift amount to balance the soil settlement caused by the gap at the tail of the shield. At the same time, the driving speed should not be too fast when the shield tunneling is crossing, and the construction speed when crossing the short pile foundation building can be controlled within 1 cm^{-2} cm/min. When passing the building, the observation frequency of settlement and tilt should be strengthened, and the driving parameters of shield should be adjusted timely according to the measured data, such as excavated quantity, pressure of soil bunker, driving speed, grouting quantity, etc.

(2) Modern home (long pile foundation)

Shield 6 passes through the north of a residential building of modern homes. Figure 2.24 shows the relationship between the excavation face and the location of the building in the process of shield tunneling. Figure 2.25a shows the settlement change curve of the measured point on the building. It can be seen that the residential building has been uplifted under the influence of shield construction. However, as the building is a pile foundation, the amount of uplift generated is well controlled, all within 3 mm. Figure 2.25b shows the deformation curve of the measuring point that is approximately a straight line on the building. It can be seen that when the shield tunneling at the side of the building, the middle part of the building will rise at first and the two ends will settle slightly. Gradually the ends of the building began to swell and the middle deformation was slight; The building as a whole bulges when the excavation face leaves the building.

According to the comparison of the measured results in Figs. 2.24 and 2.25, it is found that the structure deformation caused by shield tunneling at about 6 m away from the long-pile building is small, and it is mainly manifested as a trace of uplift, which is mainly related to the excavation rate controlled at about 98%.When adjacent to long-pile foundation buildings, the driving speed of shield tunneling can be controlled within 3–4 cm/min, the grouting pressure around 0.3 Mpa, and the grouting rate within 150–180%. When the distance of long pile building is 6 m or above, it can be considered that shield tunneling has little influence on the deformation of the structure. Only the excavation rate, grouting rate and tunneling speed can be controlled within the normal range.

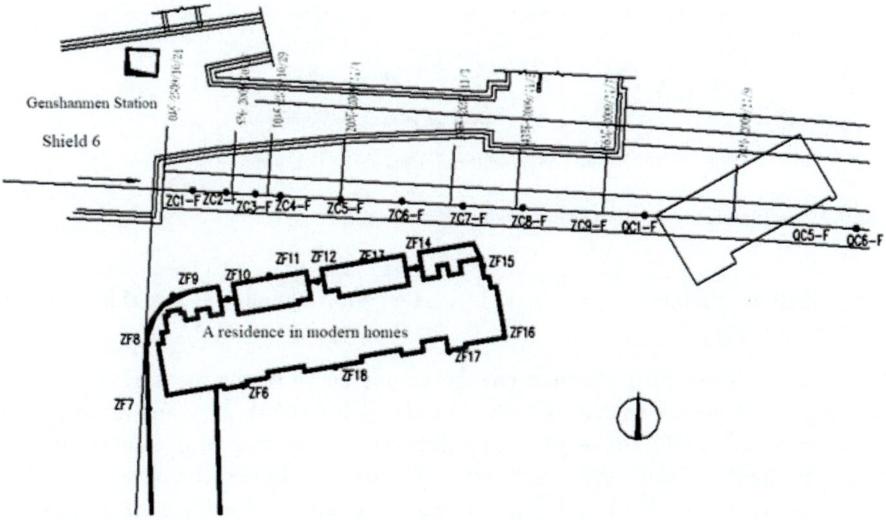

Fig. 2.24 Relation between excavation face and building location

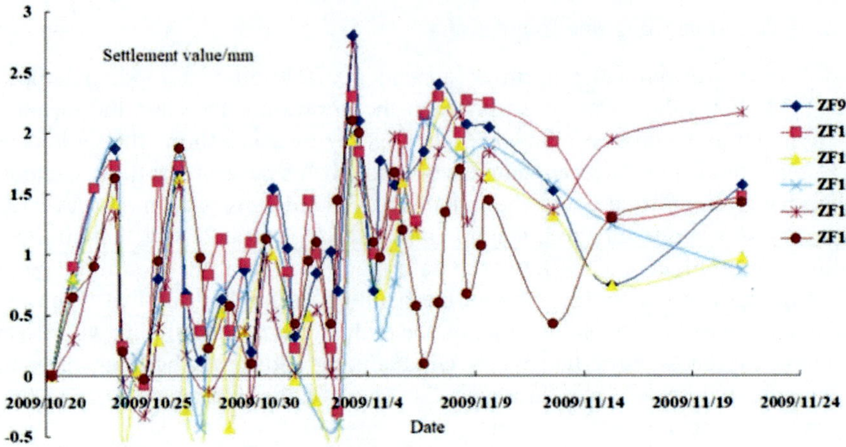

(a) Settlement history of measuring points on buildings

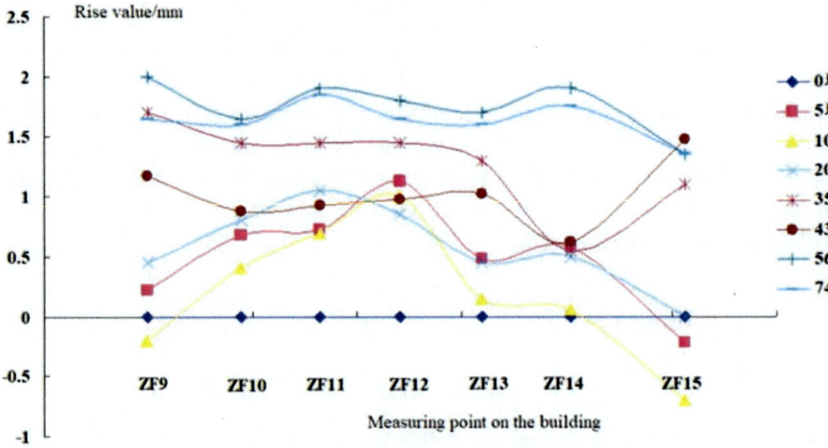

(b) The overall deformation of the building

Fig. 2.25 Settlement of modern home residential buildings

(3) Shallow foundation building near Hangzhou Wenhui Fruit Market (strip foundation)

Figure 2.26 shows the settlement change course curve of the measuring point on building A. It can be seen that when the shield incision arrives at A, all the measuring points begin to rise slightly, while the uplift of each measuring point is more obvious when the shield tail arrives at A and stays away from A. In the whole stage of uplift, the uplift amount of No.1 and No.3 measuring points is about 4 mm, and that of No.4 measuring point is about 3 mm. The overall uplift of the building is relatively uniform. When he left A shield tail, you can see, the three points immediately showed A trend of sinking, and sinking fast, when the shield tail after leaving the building

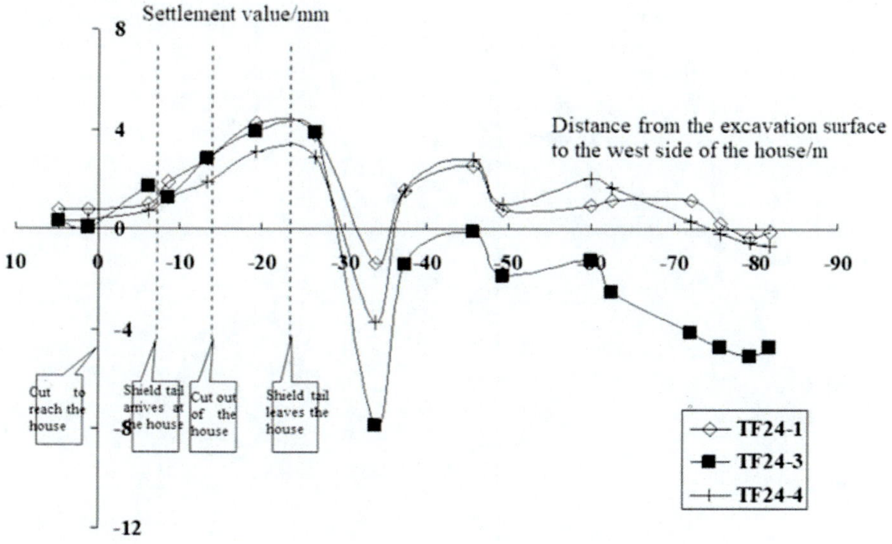

Fig. 2.26 Settlement process curve of measured point on building A

around 11 m, station reach maximum subsidence, accumulated settlement of the site of No. 3, 12 mm, no. 1 and no. 4 of the cumulative settlement of 6 mm and 7 mm respectively, as A whole building to the inclined tunnel side. After the occurrence of such land subsidence, the second grouting is adopted to control the settlement of the building. No. 1 and No. 4 points are in the state of uplift again, while No. 3 measuring point is still in the state of subsidence. The building A is inclined to the side of the tunnel.

Figure 2.27 is the settlement change course curve of the measuring points on building B. Before the shield arrives at building B, the measuring points on Building B begin to appear slight uplift. In the process when the shield arrives at B and passes through, the deformation of the other three measuring points is stable except that the measuring point No. 8 has a small settlement. When the shield is away from B and gradually away from B, the settlement of each measuring point increases gradually.

Figure 2.28a reflects the building B south three measuring points on the wall QF8, 13, 12 with the deformation of shield tunneling, can see that: (1) when the distance between shield notch and westof B is from 10.8m to 0m, measuring point 8 lifts about 1mm, No. 13 point sink is about 0.2 mm, No. 12 point down about 1 mm, this shows that building B overall tilt happens, shield squeezing soil ahead, and soil squeezing the building above, makes the B on the west side of the uplift, and as a result of the action of building overall stiffness, and further subsidence happened at 12 point; The shield penetrates the building B from 0 m to −13.2 m to the west of B. In the process, point 8 sinks a lot, about 3 mm, and point 13 sinks about 1 mm. However, point 12 almost stays the same. In the process from −13.2 to −22.8 m from the west side of B to the shield, point 8 sinks about 0.8 mm, point 13 sinks

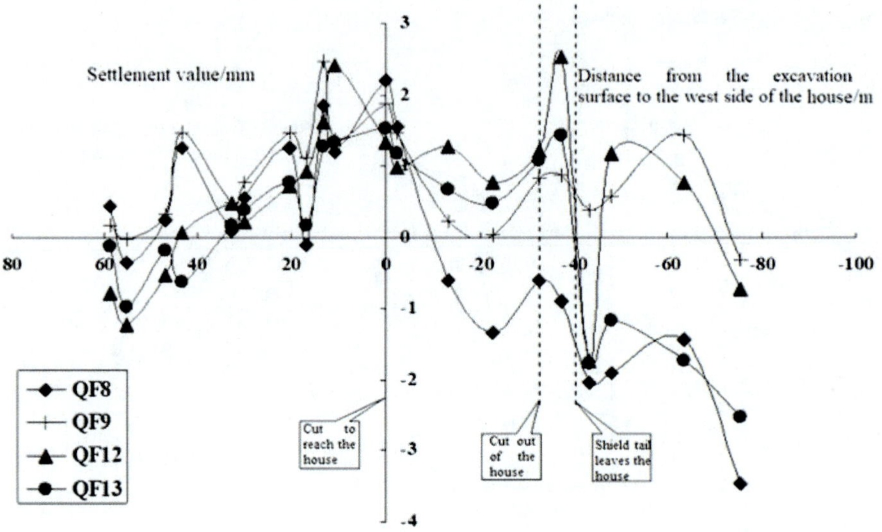

Fig. 2.27 Settlement process curve of measured point on building B

slightly, and Point 12 sinks about 0.5 mm. At this stage, the whole building B sinks, but the settlement on the east and west sides are greater than that on the middle side, so the bending degree of B is relatively large. During the distance from the shield to the west side from −22.8 m to −31.4 m (the incision left the building B 1.4 m), point 8, Point 13 and Point 12 were lifted 0.8 mm, 0.5 mm respectively. The overall uplift of building B was mainly caused by the high grouting volume and grouting pressure. In the process from the incision distance of −32.4 m from the west side of B (the incision left the building B 1.4 m) to −43.2 m (the incision left the building B 12.2 m), point 8 sank about 1.5 mm, Point 13 sank about 2.9 mm, and Point 12 sank about 3 mm. The overall settlement of the building was relatively large, and the settlement on the east side was greater than that on the west side.

Figures 2.26 and 2.27 measured results show that the same rate and the grouting rate conditions, unearthed from shallow foundation building in the process of shield tunnel crossing the pile settlement than the short buildings, buildings, uplift is greater than the amount of short pile show that the secondary grouting effect for shallow foundation building subsidence control effect is better, and the weight of the smaller buildings will produce significant lifting effect, with the literature [2]. The measured results are similar.

2.5 Summary

In this chapter, field measurements were made on four different foundation buildings in the shield tunnel interval project of Genshanmen Station (Datieguan Station) to

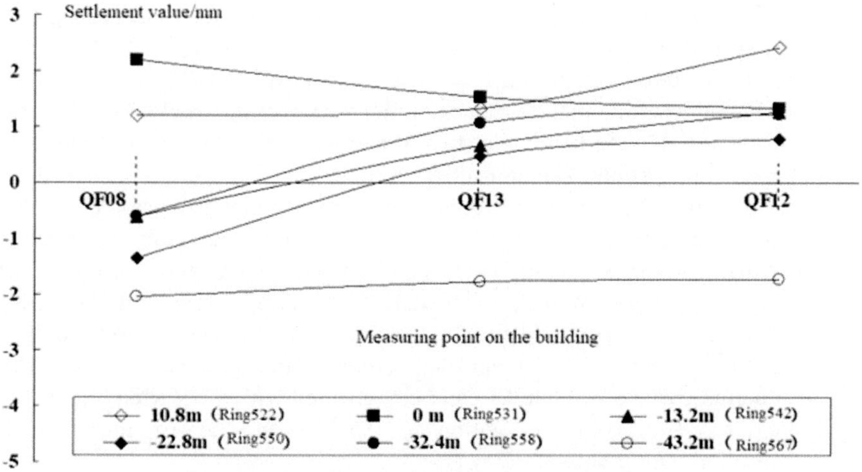

(a) Settlement changes on the south side of building B

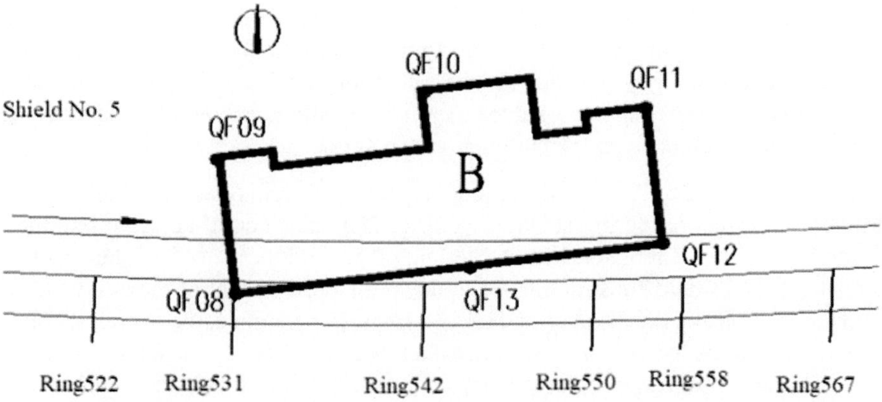

(b) The relationship between the incision of the shield machine and the location of building B

Fig. 2.28 Deformation of the house

Zhanongkou Station of Hangzhou Metro Line 1, and the influence of shield tunneling on soil deformation and adjacent different foundation buildings was studied. The field measurement results showed that:

(1) In the process of shield tunneling, deviation correction control is generally required, which will cause disturbance to the surrounding soil and further affect the deformation of adjacent buildings and their foundations. Therefore, in the construction process, the combination of automatic measurement system and manual check can be considered to control the advancing axis of the shield, so as to ensure that the tunneling axis of the shield will not deviate too much, so as to minimize the disturbance to the surrounding soil.
(2) During shield tunneling, surface uplift will occur within a certain range. The uplift of excavation surface is mainly caused by unbalance of excavation, which can be caused by underexcavation of excavation surface, excessive

support pressure and comprehensive effect of backsoil. Back-wall grouting and secondary grouting can effectively fill the gap at the tail of the shield, but if the amount of grouting is too much, the surrounding soil will also be squeezed, resulting in the surface uplift at the tail of the shield. The subsidence caused by soil disturbance and stress release during grouting is one of the main reasons for post-construction subsidence.

(3) For buildings with different foundations, shield tunneling parameters can be adjusted appropriately. When adjacent to long-pile foundation buildings, the driving speed of shield tunneling can be controlled within 3–4 cm/min, the grouting pressure around 0.3 Mpa, and the grouting rate within 150%-180%. When crossing short pile foundation and shallow foundation building, the construction speed can be controlled within 1–2 cm/min, the pressure of shield tunneling surface soil bunker can be controlled within 0.2 Mpa, the grouting pressure can be controlled within 0.3–0.5 Mpa and the grouting rate can be controlled within 160–200%.

(4) Through the shallow foundation and the short pile foundation building, shallow foundation building in the process of shield tunnel crossing the building settlement is smaller than the short pile, uplift is greater than the amount of short pile building, this is due to the short pile building is located in the hole area, the secondary grouting not timely follow-up, lead to excessive settlement, visible controlled by reasonable construction parameters on different base had a greater influence on the building deformation.

It can be seen that different foundation forms and different types of buildings show different deformation during shield tunneling. The existence of adjacent buildings restricts the movement of soil within a certain range, and the size of the deformation is related to the shield construction technology, the dead weight of the building, the foundation stiffness, the distance and other factors. For example, the overall stiffness of pile-raft foundation is generally better than that of strip and independent foundation, so it can better resist settlement, tilt and other forms of formation deformation. Therefore, it is of great significance to consider the different construction factors, structure and foundation forms in the process of shield construction to protect the buildings in the tunneling area.

References

1. Liao S, Xu J, Sun X, et al. Ground settlement patterns induced by shield tunneling and its recognition. Chinese J Underground Space Eng. 2012;8(4):777–84.
2. Xu Z, Han Q, Zheng G. Field monitoring and analysis of effects of metro tunnels under historic buildings. Chinese J Geotech Eng. 2013;35(2):364–74.

Chapter 3
Calculation of Soil Deformation Caused by Shield Tunneling

3.1 Introduction

At present, many researches have been carried out on the longitudinal and lateral displacement mechanism of soil caused by shield construction. On empirical formulas, Peck (1969) [1], who based on the long-term observation of the tunnel surface settlement trough shape, an empirical formula for predicting the normal distribution of ground settlement during tunnel construction is presented on the basis of a lot of measured data. Attewell et al. (1982) [2], who finished the calculation of longitudinal surface settlement above tunnel axis using cumulative probability formula, Liu et al. (1991) [3], who summarized the experience of tunnel construction in Shanghai and other soft land areas since 1958, and the Peck formula is revised and improved, and the longitudinal ground settlement estimation formula is put forward. In terms of theoretical formulas, Sagaseta (1987) [4, 5], who assumed that the soil mass loss is uniformly distributed along the axis, and the equivalent cylinder is used to simulate the soil mass loss, the 3-d deformation formula of soil mass is obtained. Based on the "Sagaseta formula", Verruijt and Booker (1996) [6], Loganathan and Poulos (1998) [7], Chen (2004) [8], Jiang and Zhao (2005) [9], Wei (2005) [10], Tang et al. (2010) [11] modified and improved the method, and obtained the three-dimensional semi-analytical solution of soil deformation considering the influence factors such as clearance parameters and lining deformation. However, there is an obvious defect in the above research, that is, it cannot be well explained and calculated that there may be dynamic changes of uplift and settlement at a certain distance in front of shield excavation.

Sagaseta et al. believe that the formation losses caused by tunneling are all settlement and mainly located on the excavation face. However, in fact, the formation losses of the excavation face are related to the balance of soil pressure on the excavation face, which is closely related to the excavated volume [12, 13]. Schmidt [4], Loganathan [7], Wei [10] questioned the calculation formula proposed by Sagaseta to a certain extent, and believed that the width value of land settlement trough calculated by using this theoretical formula was obviously greater than the measured value,

Z. Ding et al., *Influence of Shield Tunneling on Adjacent Structures and Control Technology*, https://doi.org/10.1007/978-981-19-1134-7_3

while the maximum value of land settlement was obviously less than the measured value. It is considered and summarized in this book that it is inappropriate to equate the soil loss caused by the soil pressure balance in the excavation surface with the soil loss caused by the gap between the shield and tail, and it is necessary to reconsider and define the concept of soil loss in the study.

In this chapter, considering the actual construction status of shield tunneling, it is considered that the soil loss caused by shield tunneling mainly includes the soil loss caused by the soil pressure balance on the excavation surface and the soil loss caused by the clearance of shield tail. Based on the convergence source method theory proposed by Sagaseta, the soil loss caused by the soil pressure balance on the excavated surface is considered to be caused by an equal amount of radial soil displacement related to the excavation rate, and the soil loss caused by the clearance of the shield tail is caused by an equal amount of radial soil displacement related to the clearance of the shield tail and grouting. Moreover, the source and sink method of Sagaseta was modified to some extent, and the formula for calculating the three-dimensional deformation of soil under semi-infinite space was derived, and the modified formula for calculating the longitudinal deformation of Sagaseta ground was given.

3.2 Principle of Huiyuan Method (Mirror Image Method)

The huiyuan method was proposed by Sagaseta [4, 5], the mirror-image method can be used to solve the problem, and the free surface problem can be considered. It can solve the displacement problem generated by voids in a homogeneous linear elastic half-space infinite body. The relevant solving steps are as follows (Fig. 3.1) [9, 14]:

(1) Firstly, the soil is assumed to be an infinite body, and the influence of the ground is ignored. Then, the problem is transformed from an infinite body in half space to an infinite internal void. Shear stress τ_0 and normal stress σ_0 will occur at the original ground, and a displacement field will occur inside the infinite body.

(2) An equal volume expansion can be assumed to exist in the interior of the infinite body at the mirror position of the original void, and the volume expansion will generate shear stress τ_0 and normal stress $-\sigma_0$ at the original ground;

(3) The normal stress generated by the above two steps at the original ground can cancel each other, and the shear stress is $2\tau_0$. In order to meet the free field boundary condition of the actual problem, the additional shear stress $2\tau_0$ inverse sign generated by the above two steps is applied to the surface of the infinite body in half space, and the displacement field generated by the stress at the ground and the following points is calculated.

(4) The sum of the displacement generated by the above three steps is the solution to the actual displacement of the problem.

The following is the specific derivation of the calculation method and process.

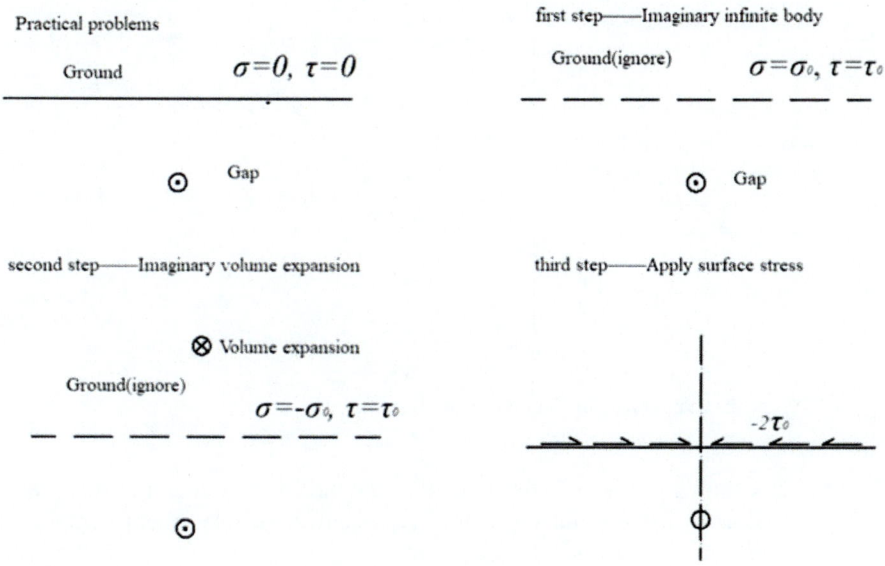

Fig. 3.1 Analysis procedure diagram of huiyuan method

As shown in Fig. 3.2, for the void inside an infinite body of half space with a radius of a, assuming that the volume does not change, the radial displacement at any point with a distance of r is

$$S_r(r) = -\frac{a}{3}\left(\frac{a}{r}\right)^2 \tag{3.1}$$

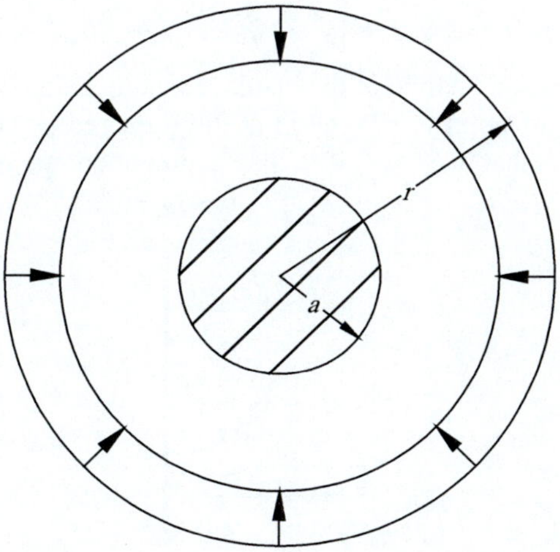

Fig. 3.2 Half infinite body void

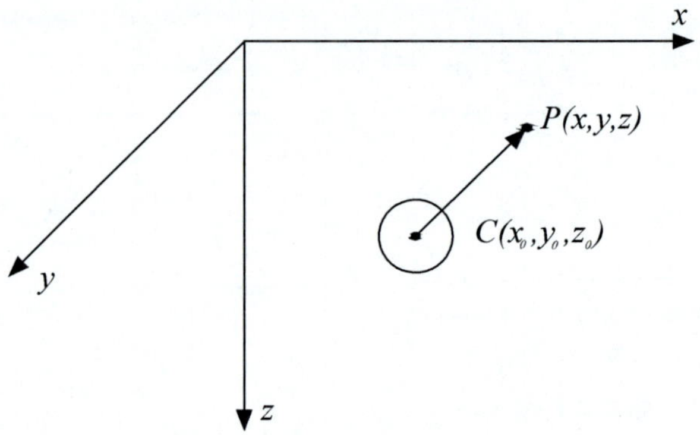

Fig. 3.3 Space diagram in rectangular coordinate system

Step 1: As shown in Fig. 3.3, in the cartesian coordinate system, the space with radius of a the point $C(x_0, y_0, z_0)$ is, and the displacement component generated at the point $P(x, y, z)$ is:

$$\left.\begin{aligned} S_{x1} &= -\frac{a^3}{3}\frac{x - x_0}{r_1^3} \\ S_{y1} &= -\frac{a^3}{3}\frac{y - y_0}{r_1^3} \\ S_{z1} &= -\frac{a^3}{3}\frac{z - z_0}{r_1^3} \end{aligned}\right\} \tag{3.2}$$

Among them, $r_1 = \left[(x - x_0)^2 + (y - y_0)^2 + (z - z_0)^2\right]^{1/2}$.

Step 2: The calculation method of the displacement generated by an equal volume expansion of point P $P(x, y, z)$ at the mirrored position $(x_0, y_0, -z_0)$ is the same as that in Step 1. The displacement components are as follows:

$$\left.\begin{aligned} S_{x2} &= \frac{a^3}{3}\frac{x - x_0}{r_2^3} \\ S_{y2} &= \frac{a^3}{3}\frac{y - y_0}{r_2^3} \\ S_{z2} &= \frac{a^3}{3}\frac{z + z_0}{r_2^3} \end{aligned}\right\} \tag{3.3}$$

Among them, $r_2 = \left[(x - x_0)^2 + (y - y_0)^2 + (z + z_0)^2\right]^{1/2}$.

Step 3: The shear strain generated in the first two steps on the surface $(x, y, 0)$ can be obtained by cauchy equation

$$\gamma_{xz} = \left(\frac{\partial(S_{x1} + S_{x2})}{\partial z} + \frac{\partial(S_{z1} + S_{z2})}{\partial x}\right)_{z=0} = -4a^3 \frac{z_0(x - x_0)}{\left[(x - x_0)^2 + (y - y_0)^2 + z_0^2\right]^{5/2}} \tag{3.4}$$

$$\gamma_{yz} = \left(\frac{\partial(S_{y1} + S_{y2})}{\partial z} + \frac{\partial(S_{z1} + S_{z2})}{\partial y}\right)_{z=0} = -4a^3 \frac{z_0(y - y_0)}{\left[(x - x_0)^2 + (y - y_0)^2 + z_0^2\right]^{5/2}} \tag{3.5}$$

According to Hooke's law, the corresponding shear stress is

$$\tau_{xz} = G\gamma_{xz}, \tau_{yz} = G\gamma_{yz} \tag{3.6}$$

where: G is the shear elastic modulus.

The distribution form of the corresponding shear stress τ_{xz} when $y = y_0$ is shown in Fig. 3.4, similarly, the distribution of the shear stress τ_{yz} can also be obtained. In order to meet the boundary conditions of the actual problem, the shear stress τ_{xz} and τ_{yz} the opposite direction are combined on the surface, and the displacement solution generated by the horizontal surface force in the known semi-infinite body is integrated (Cerruti solution) [15], the three displacement components generated in step 3 can be obtained:

$$S_{x3} = \lim_{b\to\infty} \lim_{c\to\infty} \int_{y_0-b}^{y_0+b} \int_{x_0-c}^{x_0+c} -\frac{a^3}{\pi} \frac{z_0(u - x_0)}{\left[(u - x_0)^2 + (t - y_0)^2 + z_0^2\right]5/2}$$
$$\left[\frac{1}{R} + \frac{R(R + z) - (x - u)^2}{R(R + z)^2} \cdot (1 - 2\mu) + \frac{(x - u)^2}{R^3}\right] dudt$$

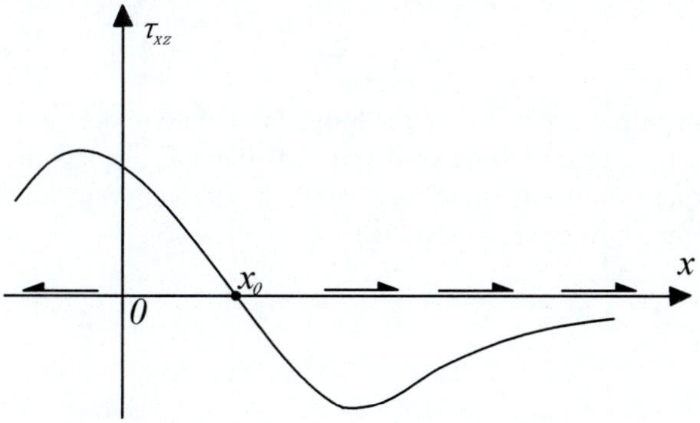

Fig. 3.4 Surface shear stress

$$+\lim_{b\to\infty}\lim_{c\to\infty}\int_{y_0-b}^{y_0+b}\int_{x_0-c}^{x_0+c}-\frac{a^3}{\pi}\cdot\frac{z_0(t-y_0)}{\left[(u-x_0)^2+(t-y_0)^2+z_0^2\right]^{5/2}}$$

$$\left[\frac{1}{R}+\frac{(y-t)^2}{R^3}+\frac{R(R+z)-(y-t)^2}{R(R+z)^2}(1-2\mu)\right]dudt \tag{3.7a}$$

$$S_{y3}=\lim_{b\to\infty}\lim_{c\to\infty}\int_{y_0-b}^{y_0+b}\int_{x_0-c}^{x_0+c}-\frac{a^3}{\pi}\frac{z_0(u-x_0)(x-u)(y-t)}{\left[(u-x_0)^2+(t-y_0)^2+z_0^2\right]^{5/2}}\left[\frac{1}{R^3}+\frac{1-2\mu}{R(R+z)^2}\right]dudt$$

$$+\lim_{b\to\infty}\lim_{c\to\infty}\int_{y_0-b}^{y_0+b}\int_{x_0-c}^{x_0+c}-\frac{a^3}{\pi}\cdot\frac{z_0(t-y_0)(x-u)(y-t)}{\left[(u-x_0)^2+(t-y_0)^2+z_0^2\right]^{5/2}}\left[\frac{1}{R^3}+\frac{1-2\mu}{R(R+z)^2}\right]dudt \tag{3.7b}$$

$$S_{z3}=\lim_{b\to\infty}\lim_{c\to\infty}\int_{y_0-b}^{y_0+b}\int_{x_0-c}^{x_0+c}-\frac{a^3}{\pi}\frac{z_0(u-x_0)(x-u)}{\left[(u-x_0)^2+(t-y_0)^2+z_0^2\right]^{5/2}}\left[\frac{z}{R}+\frac{1-2\mu}{R(R+z)}\right]dudt$$

$$+\lim_{b\to\infty}\lim_{c\to\infty}\int_{y_0-b}^{y_0+b}\int_{x_0-c}^{x_0+c}-\frac{a^3}{\pi}\cdot\frac{z_0(t-y_0)(y-t)}{\left[(u-x_0)^2+(t-y_0)^2+z_0^2\right]^{5/2}}\left[\frac{z}{R}+\frac{1-2\mu}{R(R+z)}\right]dudt \tag{3.7c}$$

where: μ is Poisson's ratio, $R=\left[(x-u)^2+(y-t)^2+z^2\right]^{1/2}$.

Step 4: The sum of the displacement generated by the above three steps is the displacement solution of the actual demand solution, which is in a semi-infinite body (x_0, y_0, z_0) the total displacement generated by the gap with radius a at point (x, y, z) is:

$$S_x=S_{x1}+S_{x2}+S_{x3},\quad S_y=S_{y1}+S_{y2}+S_{y3},\quad S_z=S_{z1}+S_{z2}+S_{z3} \tag{3.8}$$

The displacement generated by the gap per unit volume is divided by the volume $4\pi a^3/3$ on the basis of the above formula, i.e.

$$s_x=\frac{S_x}{\frac{4}{3}\pi a^3},\quad s_y=\frac{S_y}{\frac{4}{3}\pi a^3},\quad s_z=\frac{S_z}{\frac{4}{3}\pi a^3} \tag{3.9}$$

The above steps 1–4 are based on the huiyuan method to derive the soil displacement generated by a unit volume void space. If the soil displacement caused by a unit volume expansion needs to be calculated, the derivation process is completely consistent, but the signs are reversed [16].

3.3 Huiyuan Method Considering Influencing Factors of Construction

If the late consolidation deformation of the surface is not considered, the soil deformation caused by the tunneling of the shield with soil pressure balance mainly includes the deformation caused by the unbalance of soil pressure on the excavation surface and the deformation caused by the release of the shield tail and grouting. It can be considered that the excavation rate at the excavated surface of shield is 100%, that is, when the excavated surface is in the excavated equilibrium state, no uplift and subsidence phenomenon will occur at the excavated surface area [13]. If the excavation rate is more than 100%, the excavation face is in the state of over excavation, and the soil will sink. If the excavation rate is less than 100%, the excavation face is in the state of compaction, and the soil mass will be uplifted. At the same time, in the process of shield tunneling, annular gap will inevitably occur in the shield tail, which will result in soil subsidence. In order to avoid excessive displacement of soil, the grouting hole of shield tail should be timely grouting to reduce the amount of soil settlement. If the gap of shield tail can not be effectively grouting, the soil deformation still appears subsidence phenomenon; If the gap of the shield tail is filled with slurry or even overgrouting, the deformation of soil at the shield tail will appear uplift phenomenon. Based on this, huiyuan method is adopted in this book to calculate the deformation of soil caused by unbalance of excavation face, gap of shield tail and synchronous grouting respectively. Excavation rate, grouting rate and other construction factors can be considered to make the calculation results more in line with the actual project.

3.3.1 Calculation of Soil Deformation Caused by Unbalance of Excavation Face

From Yan [12], it is considered that the amount of soil discharge in shield construction has a direct influence on the stability of shield excavation face, and controlling the amount of soil discharge is an important measure to control surface deformation. From Wang [13], it is considered that the excavation rate of the earth-pressure balanced shield is an important parameter to control the balance during the shield propulsion, which determines the size of the formation loss and the size of the uplift or settlement in front. When the cutting volume of shield tunneling is equal to the volume of spiral excavator, the shield tunneling is in a state of equilibrium, which is more meaningful for the control of the shield on soil pressure balance than the equilibrium of the contact pressure and water and soil pressure between the shield and the front soil. It can be seen that the conversion calculation of excavated quantity and soil loss in the excavated surface is more in line with the construction practice, and the concept is also clearer.

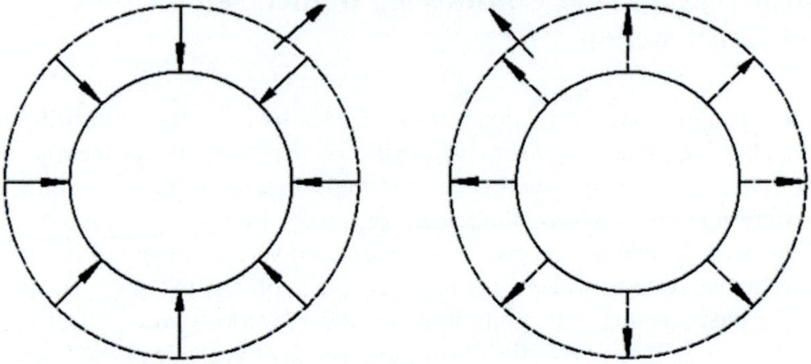

Fig. 3.5 Soil movement mode diagram of excavation surface

Therefore, the huiyuan method is introduced into the excavation surface to calculate the soil deformation caused by unbalance. The calculation process is as follows:

The soil loss caused by unbalance of excavation face is related to the excavated quantity. It is assumed that the soil is an isotropic elastic medium, and the excavated surface soil shows a circular equivalent radial displacement movement pattern due to unbalance of excavated soil, as shown in Fig. 3.5. When shield excavation surface in the unearthed rate is greater (smaller) than 100%, machine surrounding soils produce inward (outside) side of the uniform radial displacement, it is by the unit length "positive (negative) of soil loss" caused by the displacement of soil produced the size of the area and unearthed rate, the actual rate per unit length of unearthed from soil loss computation formula is as follows:

$$V_{loss}^{1} = \pi R^2(\xi - 1) \tag{3.10}$$

where: R is the outer diameter of shield tunneling machine, and ξ is the excavation rate.

The excavation rate ξ can be calculated from the excavation record at the construction site as follows:

$$\xi = V_c/\pi R^2 l_G \tag{3.11}$$

where: V_c is the output of each ring, is the automatic recording parameter of the shield machine, l_G is the length of each ring piece.

If the excavated volume of shield tunneling machine cannot be recorded, the excavated rate ξ can also be calculated according to the following empirical formula [13]:

$$\xi = \frac{4\eta k k_c Q}{\pi \gamma_0 D^2} \times \left(\frac{N}{v}\right) \tag{3.12}$$

where: η is the excavation efficiency of the screw machine, which is related to soil properties and rotation speed. k is a parameter to convert volume to weight, which is related to the nature of soil layer; k_c is a parameter considering the effective excavated ratio when adding material weight; Q is the volume of unearthed from the screw machine; N is the speed of the screw; γ_0 is the natural bulk density of soil; D is spiral diameter; v is the propulsion speed of shield tunneling machine.

When the excavation rate at the shield excavation face is 100%, the soil displacement area is 0, which means that the soil loss is 0. The soil mass at the excavation face will not heave and sink. The displacement clearance caused by the imbalance of the excavation face is calculated as follows:

$$g_1 = R \times \left(\sqrt{\xi} - 1\right) \tag{3.13}$$

Displacement of excavation surface in the soil or size can be based on unit volume gap of expansive soil displacement produced by the integral, as shown in Fig. 3.6, assumes that the depth of tunnel excavation face center is h, since the point (0, 0, h) along the x axis is the direction started excavating, advance to the location as shown (0, l, h), thickness for clearance is g_1. It can be seen that the soil loss is the space between two cylinders with the same length and different radius, i.e. $V = V_1 - V_2$. Where, when the excavation rate is greater than 100%, the radius of the outer circle is R and the radius of the inner circle is $R \times \left(2 - \sqrt{\xi}\right)$. When the excavation rate is less than 100%, the radius of the outer circle is $R \times \left(2 - \sqrt{\xi}\right)$, and the radius of the inner circle is R.

The soil loss and displacement caused by the excavation of the excavation face can be obtained by the following integration:

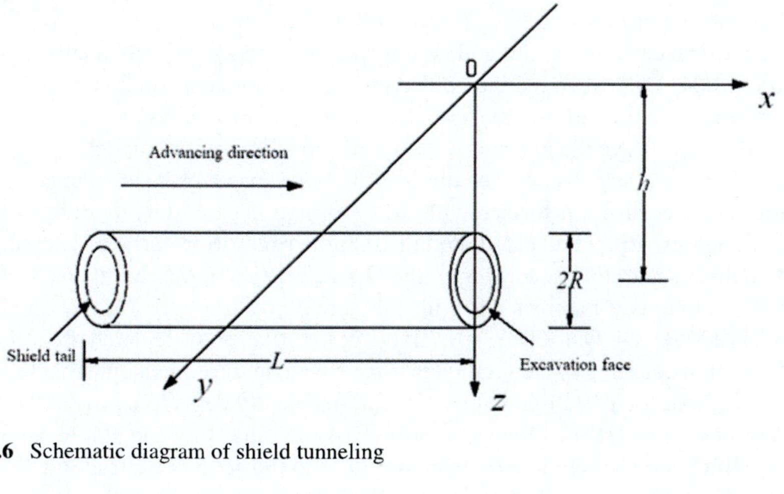

Fig. 3.6 Schematic diagram of shield tunneling

$$U^1 = \iiint_V s_x(x, y, z)\mathrm{d}x\mathrm{d}y\mathrm{d}z = \iiint_{V_1 - V_2} s_x(x, y, z)\mathrm{d}x\mathrm{d}y\mathrm{d}z \tag{3.14a}$$

$$V^1 = \iiint_V s_y(x, y, z)\mathrm{d}x\mathrm{d}y\mathrm{d}z = \iiint_{V_1 - V_2} s_y(x, y, z)\mathrm{d}x\mathrm{d}y\mathrm{d}z \tag{3.14b}$$

$$W^1 = \iiint_V s_z(x, y, z)\mathrm{d}x\mathrm{d}y\mathrm{d}z = \iiint_{V_1 - V_2} s_z(x, y, z)\mathrm{d}x\mathrm{d}y\mathrm{d}z \tag{3.14c}$$

Type in the U^1 is the displacement in the x direction; V^1 is the displacement in the y direction; W^1 is the displacement in the z direction.

3.3.2 *Calculation of Soil Deformation Caused by Gap of Shield Tail*

According to the characteristics of shield tunneling, it can be considered that the movement of soil in the excavation face, the penetration of soil into the shield tail clearance and the grouting of the shield tail are the main causes of surface deformation, among which the amount of soil penetration into the shield tail clearance is the most important factor determining the amount of soil settlement. As the diameter of the cutter plate of the shield tunneling machine is larger than the outer diameter of the lining pipe piece, after the pipe piece is assembled and detached at the tail of the shield, an annular gap is formed between the pipe piece and the soil, which is referred to as the tail gap of the shield, as shown in Fig. 3.7. If the shield tail gap is not filled in time, it is bound to cause the soil to sink, and then cause the settlement of adjacent surface construction and structures. Lee (1992) [17], Loganathan (1998) [7], Park (2004) [18] proposed the equivalent soil clearance parameter g, and calculated and analyzed the three-dimensional soil deformation caused by the clearance at the shield tail by using the non-equivalent radial soil movement model, as shown in Fig. 3.8. However, they all confused the deformation caused by the unbalance of earth pressure on the excavation surface with the deformation caused by the release of the shield tail and grouting, and calculated them at the excavation surface. They did not calculate them separately according to the characteristics of the shield construction, which was obviously inconsistent with the actual construction. And Zhao (2004) [16], Wei (2005) [10], Qi (2007) [19], Zhu (2011) [20] et al. Calculated the influence of deformation caused by shield tail clearance based on the equivalent soil clearance parameter g proposed by Loganathan [7], and repeatedly considered the thrust force of shield machine and the change of soil pressure caused by the shield tunneling surface, which are also obviously inconsistent with the actual construction.

Zhao (2004) [16], Wei (2005) [10], Qi (2007) [19], Zhu (2011) [20] who based on the equivalent soil clearance parameter G proposed by Loganathan and Poulos, the

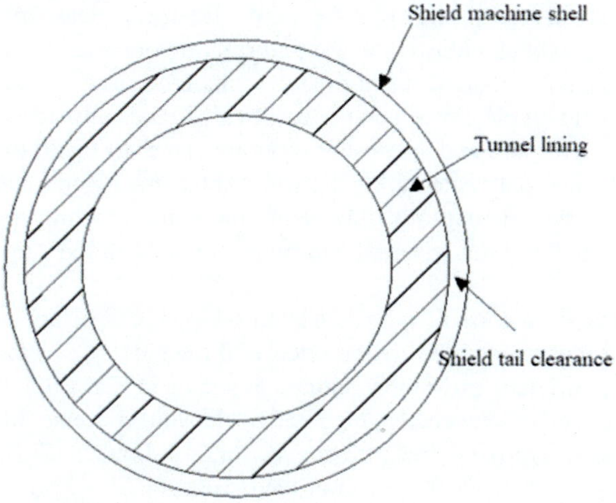

Fig. 3.7 Schematic diagram of shield tail clearance

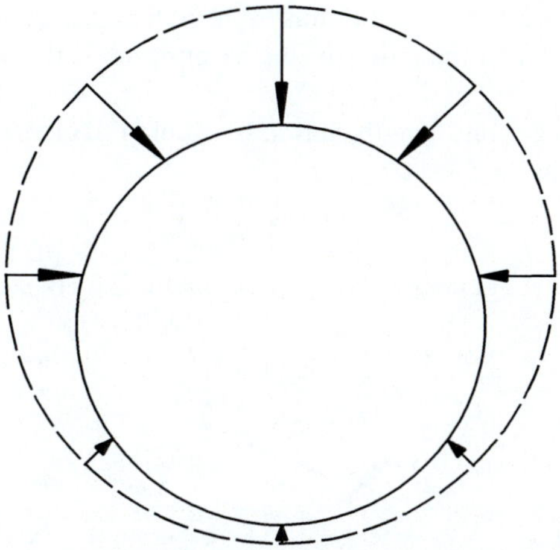

Fig. 3.8 Is a schematic diagram of movement mode of non-equivalent radial soil mass

soil movement model caused by shield tail caving is non-equivalent radial soil movement model. Wei [10] considered that the maximum land settlement value generated directly above the tunnel axis in the non-equal radial soil mass movement mode will be larger than that in the uniform radial displacement mode, which solves the problem that the maximum land settlement value should be significantly smaller than the measured value. In this book, it is believed that considering the non-equivalent radial soil movement mode is worth discussing. Firstly, the clearance of shield tail is an instantaneous process, and stress release is required for the annular soil clearance formed by shield tail, while stress release is required for both upper and lower soil, which is obviously inappropriate to consider only the stress release of upper soil [21]. Secondly, Wei Gang et al. believe that the deformation obtained by using the equivalent uniform radial soil displacement model is small, because it is calculated on the basis of equivalent soil clearance parameter g, which takes into account the excavation face support pressure and shield-tail grouting force. In fact, other factors need not be included when only calculating the soil displacement generated by the shield tail clearance, so the calculation results using the equal amount of uniform radial displacement model will not be small, but more in line with the actual construction situation.

It is still assumed that the soil mass is an isotropic elastic medium, and the displacement of the soil mass generated by the release of the shield tail presents a circular equivalent radial soil mass movement pattern, as shown in Fig. 3.9. It can be obtained by integrating the soil displacement generated by the void space per unit volume. The derivation process is similar to the previous section. As shown in Fig. 3.6, we assume that the central section depth at the end of the tunnel shield is h, from point $(0, -L, h)$, driving along the positive direction of x axis, and advancing to the position $(0, l{-}L, h)$, as shown in the figure, and the clearance thickness is $g_2 = R - r$. The soil loss is still composed of gaps between two cylinders of the same length and different radii, i.e. $V' = V_1' - V_2'$. Where, the radius of the outer circle is R, the outer diameter of the shield machine, and the radius of the inner circle is the outer diameter of the lining r.

Then the soil loss per unit length caused by shield tail clearance is

$$V_{loss}^2 = \pi\left(R^2 - r^2\right) \tag{3.15}$$

The soil displacement generated by it can be obtained by the following integration:

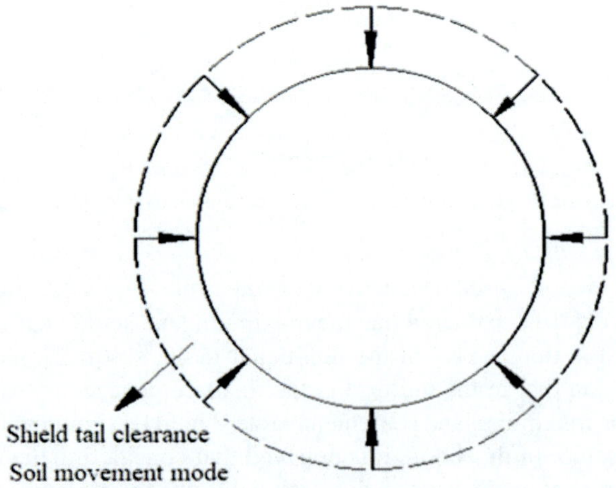

Fig. 3.9 Diagram of soil movement mode in the shield tail clearance

$$U^2 = \iiint_{V'} s_x(x, y, z)dxdydz = \iiint_{V_1'-V_2'} s_x(x, y, z)dxdydz \quad (3.16a)$$

$$V^2 = \iiint_{V'} s_y(x, y, z)dxdydz = \iiint_{V_1'-V_2'} s_y(x, y, z)dxdydz \quad (3.16b)$$

$$W^2 = \iiint_{V'} s_z(x, y, z)dxdydz = \iiint_{V_1'-V_2'} s_z(x, y, z)dxdydz \quad (3.16c)$$

In the type: U^2 is the displacement in the x direction; V^2 is the displacement in the y direction; W^2 is the displacement in the z direction.

3.3.3 Calculation of Soil Deformation Caused by Grouting at Shield Tail

According to the analysis in the above section, when the shield tail comes out of the pipe, an annular gap will be formed between the pipe and the soil. If the clearance is not filled in time, it will lead to a large settlement of the soil within a certain range above. In order to effectively control this part of settlement, shield tail synchronous grouting is often used. The shield machine synchronously infuses sufficient grout into the gap through the embedded grouting pipe to fill the annular gap of the shield tail. At the same time, the slurry filled with gaps can provide a certain pressure to support the soil above and prevent the soil above from collapsing after the shield tail is empty and causing a large instantaneous settlement [22]. Zhao (2004) [16], who considered that the grouting pressure of the grouting hole above the shield tail is very large, while that of the grouting hole below is very small, and the grouting volume outside the lining is mainly concentrated on the top of the tunnel, while the grouting volume at the bottom is relatively small. But Bezuije (2004) [23] monitored the change of slurry pressure with tunneling time in Sophia tunnel on site, and it was found that the slurry pressure distribution was more uniform after grouting was completed, and the slurry pressure showed the form of small on top and large on the bottom. In this book, it is believed that in the actual grouting construction, the amount of grouting outside the lining does not necessarily show a distribution of large amount of grouting up and small amount of grouting down or large amount of grouting up and small amount of grouting down. The actual filling situation of shield tail grouting is shown in Fig. 3.10.

According to the distribution characteristics of grouting amount, considering that the distribution of grouting amount is greatly affected by soil quality, grout material, grouting pressure and other construction technologies, this book assumes that the cross section formed after grouting is similar to the annular distribution of the shield

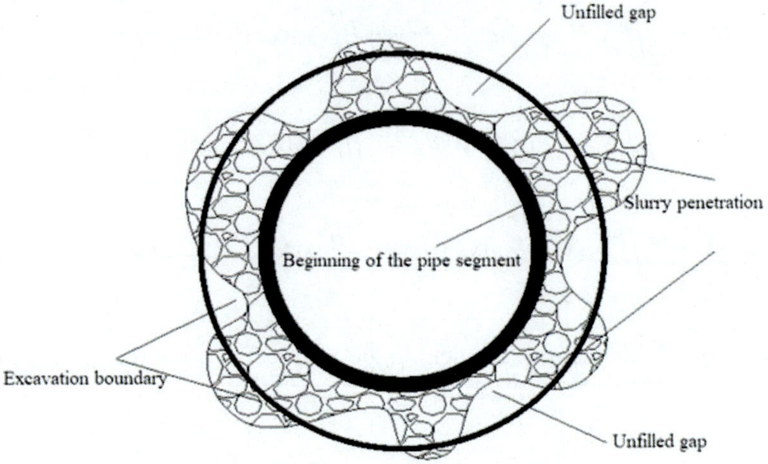

Fig. 3.10 A schematic diagram of grouting filling at the tail of shield

tail clearance, but still presents an ideal uniform annular, and the grouting process can be considered as the reverse process generated by the shield tail clearance [16]. But the displacement volume of soil caused by grouting is not equal to the volume of grouting in the actual construction, because the slurry will be lost in the actual grouting process [19]. Therefore, it can be considered that the displacement generated by grouting is volume expansion and negative soil loss, and its size is related to the grouting amount, grouting rate and slurry loss rate. The calculation formula of unit length is as follows:

$$V_{loss}^{3} = (1 - \eta) \times \lambda\left(\pi R^2 - \pi r^2\right) \tag{3.17}$$

where: R is the outer diameter of shield tunneling machine; r is the outer diameter of lining; λ is the grouting rate; η is the grouting loss coefficient.

The grouting rate λ can be calculated by the following formula according to the grouting volume record on the construction site:

$$\lambda = V_z / \pi R^2 l_G \tag{3.18}$$

where: V_z is the amount of grouting in each ring, is the parameters recorded during shield tunneling process, and l_G is the length of each ring piece.

Slurry loss coefficient is generally closely related to soil porosity and permeability coefficient, tunnel overbreak, grouting pressure, slurry pipeline length and other factors [19, 24], the calculation formula is as follows:

$$\eta = \alpha_1 + \alpha_2 + \alpha_3 + \alpha_4 \tag{3.19}$$

where, α_1 is the compression density loss coefficient, which is related to the material properties of the slurry, and refers to the phenomenon that the density of the slurry increases and the volume decreases. The change range is between 0.05 and 0.15, usually 0.1. α_2 is the soil loss coefficient, which is related to the soil properties. For soils with large porosity and permeability, the slurry loss is also larger. Generally, 0.35 can be used for soft soil. α_3 is the transport loss coefficient, which refers to the loss to a certain extent due to a small amount of slurry remaining in the tube during slurry transport. For tunnels with a general length of 1000 m or less, the transport loss coefficient can be 0.1; α_4 is the overcut loss coefficient, the slurry should be filled with the overcut gap to cause the slurry loss, generally 0.05.

It is still assumed that the soil is an isotropic elastic medium, and the volume expansion generated by shield tail grouting presents a circular equivalent radial soil movement pattern, as shown in Fig. 3.11. The derivation process is similar to that in the previous section. As shown in Fig. 3.6, assume that the central depth at the section of tunnel shield tail is, from point (0, –*L, h*), driving along the positive direction of *x*-axis, and advancing to the position (0, *l–L, h*) as shown in the figure, and the clearance thickness is $g_3 = \sqrt{\frac{V_{loss}^3}{\pi} + r^2} - r$. The soil loss is still composed of gaps between two cylinders of the same length and different radii, i.e. $V'' = V_1'' - V_2''$. Where, the radius of the outer circle is $\sqrt{\frac{V_{loss}^3}{\pi} + r^2}$, and the radius of the inner circle is the lining outer diameter is r.

The soil displacement generated by shield tail grouting can be obtained by the following integral:

$$U^3 = \iiint\limits_{V''} s_x(x, y, z)dxdydz = \iiint\limits_{V_1''-V_2''} s_x(x, y, z)dxdydz \quad (3.20a)$$

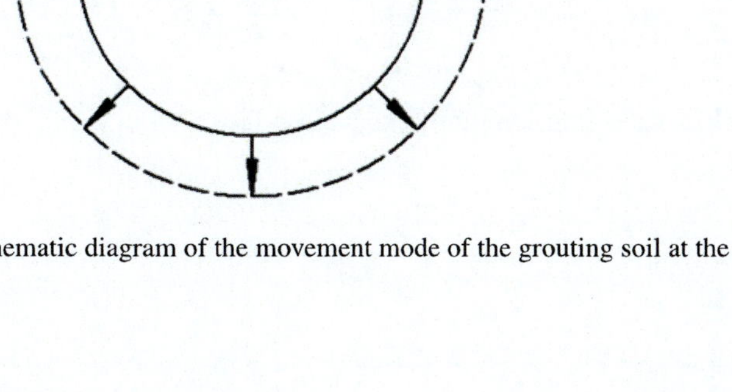

Fig. 3.11 A schematic diagram of the movement mode of the grouting soil at the end of shield

$$V^3 = \iiint\limits_{V''} s_y(x, y, z)dxdydz = \iiint\limits_{V_1''-V_2''} s_y(x, y, z)dxdydz \quad (3.20b)$$

$$W^3 = \iiint\limits_{V''} s_z(x, y, z)dxdydz = \iiint\limits_{V_1''-V_2''} s_z(x, y, z)dxdydz \quad (3.20c)$$

In the type U^3 is the displacement in the x direction; V^3 is the displacement in the y direction; W^3 is the displacement in the z direction.

To sum up, the total deformation of soil caused by the unbalance of soil pressure on the excavated surface, the release of the shield tail and the grouting during the tunneling process of the shield is as follows:

$$U = U^1 + U^2 - U^3, \quad V = V^1 + V^2 - V^3, \quad W = W^1 + W^2 - W^3 \quad (3.21)$$

where, U is the displacement in the x direction; V is the displacement in the y direction; W is the displacement in the z direction.

3.4 Improved Sagaseta Calculation Formula

3.4.1 Improvement of Sagaseta Formula

According to the analysis in the previous section, factors such as unbalance of excavation surface and empty of shield tail during shield construction will lead to soil deformation, which will lead to excessive ground settlement or uplift, and thus affect the surrounding buildings. The ground deformation caused by shield tunneling deserves more attention. Based on Formula (3.14), Sagaseta used the convergence source method to obtain the analytic solution of ground deformation caused by soil loss under plane conditions. The calculation formula is as follows:

$$\left.\begin{aligned} S_{x0} &= -\frac{v}{2\pi} \times \frac{y}{y^2+h^2}\left[1 - \frac{x}{\sqrt{x^2+y^2+h^2}}\right] \\ S_{y0} &= \frac{v}{2\pi} \times \frac{1}{\sqrt{x^2+y^2+h^2}} \\ S_{z0} &= \frac{v}{2\pi}\frac{h}{y^2+h^2}\left[1 - \frac{x}{\sqrt{x^2+h^2+y^2}}\right] \end{aligned}\right\} \quad (3.22)$$

where: v is soil loss per unit length of excavation face.

Equation (3.22) is more easily applied in engineering than the three-dimensional solution of soil deformation (3.14). However, it only considers the soil loss occurring in the excavation face, and does not consider the influence of shield construction factors. Moreover, the calculated surface deformation cannot consider the influence of uplift. Therefore, this book refers to the ground deformation analytical solution derived by Sagaseta based on the confluent method, and combined with Eq. (3.21), the vertical ground deformation solution under the consideration of construction influence factors can be deduced as follows:

$$\begin{aligned} S_{z0} = & \frac{V_{loss1}}{2\pi}\frac{h}{y^2+h^2}\left[1-\frac{x}{\sqrt{x^2+y^2+h^2}}\right] \\ & + \frac{V_{loss2}}{2\pi}\frac{h}{y^2+h^2}\left[1-\frac{x+L}{\sqrt{(x+L)^2+y^2+h^2}}\right] \\ & - \frac{V_{loss3}}{2\pi}\frac{h}{y^2+h^2}\left[1-\frac{x+L}{\sqrt{(x+L)^2+y^2+h^2}}\right] \end{aligned} \tag{3.23}$$

In the formula, V_{loss1}, V_{loss2} and V_{loss3} are respectively soil loss per unit length caused by excavation surface excavation, soil loss per unit length caused by shield tail clearance and soil loss per unit length caused by grouting.

In the process of shield tunneling, if the excavation rate is 100%, that is, when the excavation remains balanced, the formula (3.23) can be converted into the following formula:

$$S_{z0} = \frac{(V_{loss2}-V_{loss3})h}{2\pi\left(y^2+h^2\right)} \times \left[1-\frac{x+L}{\sqrt{(x+L)^2+y^2+h^2}}\right] \tag{3.24}$$

The vertical solution in Eq. (3.24) is similar to that in Eq. (3.22), but the concept of soil loss is more clear. Its location is at the tail of the shield instead of the excavation face, and its size is determined by the clearance of the tail of the shield and grouting under the condition that the excavation face remains balanced, which is more in line with the actual construction conditions.

For Eq. (3.23), when $y = 0$, the calculation formula of longitudinal ground deformation above the tunnel axis is as follows:

$$\begin{aligned} S_{z0} = & \frac{V_{loss1}}{2\pi h}\left[1-\frac{x}{\sqrt{x^2+h^2}}\right] + \frac{V_{loss2}}{2\pi h}\left[1-\frac{x+L}{\sqrt{(x+L)^2+h^2}}\right] \\ & - \frac{V_{loss3}}{2\pi h}\left[1-\frac{x+L}{\sqrt{(x+L)^2+h^2}}\right] \end{aligned} \tag{3.25}$$

The following examples are used to verify the improved Sagaseta formula (3.24) and (3.25), so as to show the correctness and applicability of the improved formula.

3.4.2 Example Verification

(1) Example analysis i

Example 1: Based on the field measurements in this book, the diameter of the shield tunneling machine is 6.34 m, the captain is 8.5 m, the inner and outer diameters of the lining are 5.50 m and 6.20 m respectively, the width is 1.2 m, and the depth of the tunnel axis is about 14 m. See Chap. 2 for the parameters related to the soil layer.

The above analysis shows that the grouting rate of construction parameters is 183%, the grouting loss rate is 65% and the excavation rate is 98.9% . The calculation results are shown in Fig. 3.12. It can be seen from the figure that the theory is consistent with the measured results. In the theory of the longitudinal settlement curve of shield excavation front of 20 m have been uplift, the uplift is mainly by the excavated volume is less than the volume of cutting soil, soil chamber pressure is greater than the static earth pressure in front of the excavation, and make the front shield and top soil caused by extrusion, behind the shield of subsidence mainly grouting amount is less than the gap segment after. There is settlement less than 0.5 mm in the front 10–20 m range of the measured curve, which may be caused by road traffic load.

(2) Example analysis ii

Example 2: For an interval tunnel project of Hangzhou Metro Line 2, the diameter of shield tunneling machine is 6.34 m, the captain is 8.5 m, the lining inner and outer diameters are 5.50 m and 6.20 m respectively, the width is 1.2 m, and the overlaying thickness of the tunnel top is 9.0 m. The soil layer from top to bottom is as follows:

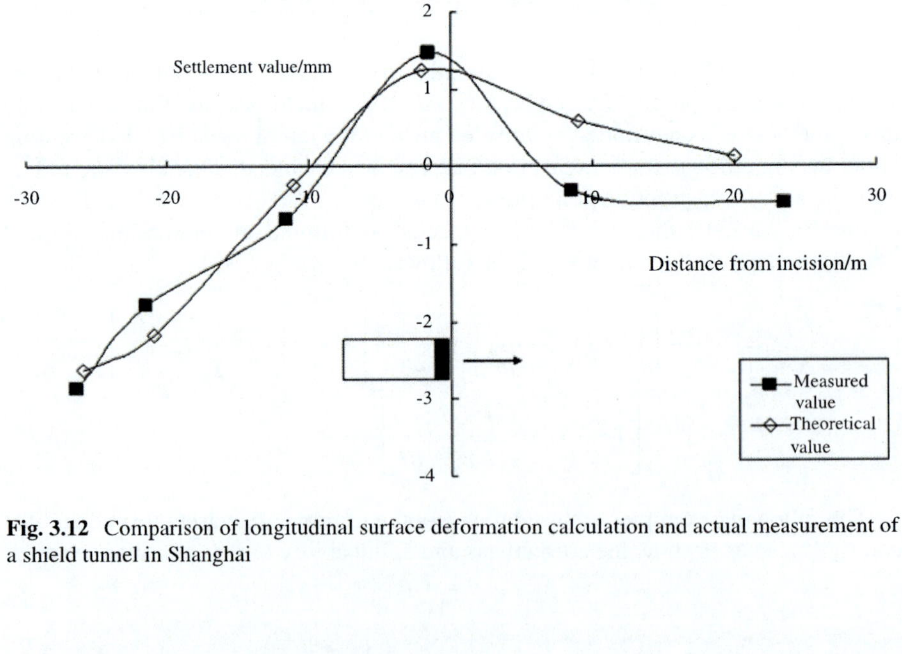

Fig. 3.12 Comparison of longitudinal surface deformation calculation and actual measurement of a shield tunnel in Shanghai

① Sandy silty soil; ② Silty silty clay; ③ Sandy silty soil with silt silty clay; ④ Sandy silt with silt; ⑤ Silty sand with sandy silty soil; ⑥ Sandy silty soil. The tunnel mainly passes through ④ sandy silt mixed with silt and ⑤ sandy silt mixed with sandy silt [25].

According to document [25], shield construction parameters of the grouting rate is between 150 and 180%, grouting attrition rate between 60 and 70%, the rate is between 95 and 100%, unearthed from the calculation of the book when grouting rate was 163%, grouting loss was 65%, the unearthed at a rate of 98.6%, the results shown in Fig. 3.13, the figure is tallies with the theory and the measured settlement curve, shield machine to produce a certain amount of uplift in front, is mainly caused by the excavation face unearthed imbalance, but the maximum uplift position slightly deviation, this is mainly theoretical calculation value of not considering the influence of deflection and the friction force of shield machine.

(3) Calculation example analysis iii

Example 3: The shield tunneling section of the second phase of the Mingzhu Line project of Shanghai Rail transit is located between the Linping Road station and the Piaoyang Road Station. A full-section cutting earth-pressure balancing shield machine is used for construction. The diameter of the shield machine is 6.34 m, the length is 6 m, the inner diameter and outer diameter of the lining are 5.50 m and 6.20 m respectively, and the lining width is 1.0 m. The tunnel was buried 9.1 m deep and mainly passed through layers 3–1 clayey silt and 3–2 sandy silt. The other soil layers are the first layer of artificial fill, the fourth layer of silt clay [26].

When calculating the longitudinal surface deformation, the construction parameters of shield tunneling are taken according to the actual working conditions: grouting rate is 200%, grouting loss is 65%, and excavation rate is 100%. See Fig. 3.14 for the calculation results. The results show that the theoretical value is in good agreement

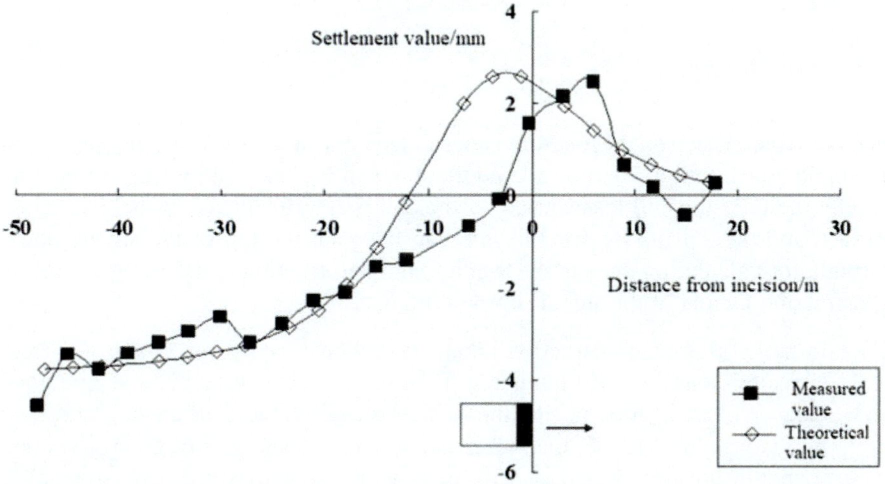

Fig. 3.13 Comparison of longitudinal surface deformation calculation and actual measurement of a shield tunnel in Hangzhou

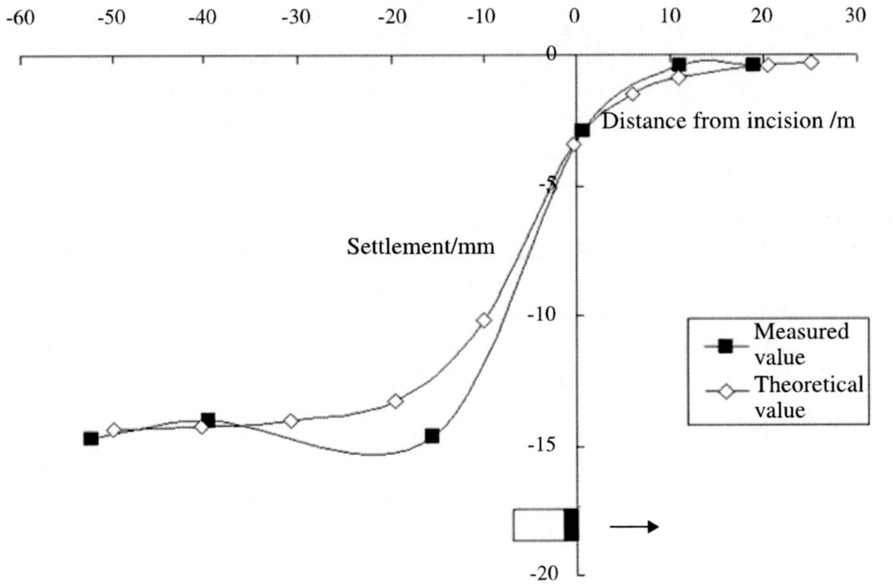

Fig. 3.14 Comparison of longitudinal surface deformation calculation and actual measurement of a shield tunnel in Hangzhou

with the measured value, and the excavated value is relatively balanced under this working condition. There is no uplift in front of the shield machine, and the shield tail after passing through produces a large settlement, reaching the corresponding peak value.

3.5 Summary

In this chapter, based on the measured data and in view of the soil deformation caused by shield construction, it is considered that the soil loss caused by shield tunneling mainly includes the soil loss caused by the soil pressure balance of the excavation surface and the soil loss caused by the gap between the tail of the shield, and a formula for calculating the surface longitudinal deformation considering the actual construction factors of the shield is proposed.

(1) In soft soil area, construction parameters should be set reasonably for earth pressure balance shield tunneling. First of all, it is necessary to keep a good balance of excavation rate on the shield excavation face. When the excavation rate is less than 100%, the soil mass before excavation will be uplifted, but it should not be lower than 95% in general, so as to avoid excessive uplift. Considering the slurry loss during the grouting of the shield tail, the grouting

rate is generally more than 100% to avoid the large surface settlement caused by the gap of the shield tail.

(2) In Sagaseta huiyuan method based on the theory, the excavation surface unearthed balance caused by soil loss considered by the associated with the unearthed rate equivalent uniform radial displacement of soil, soil damage shield tail clearance is related by the shield tail escapes and grouting amount evenly radial displacement of soil, soil and semi-infinite space is deduced and the calculation formula of deformation, and modified Sagaseta ground vertical deformation calculation formula is given, and the measured data is analyzed.

Considering the influence of shield construction technology, this book deduces the modified formula of surface and longitudinal deformation, which is in good agreement with the measured data. However, in the actual construction, secondary grouting is needed to avoid large surface settlement, which is difficult to be reflected in the calculation formula of longitudinal deformation, so further research and analysis are needed.

References

1. Peck RB. Deep excavations and tunneling in soft ground. In: Proceeding of 7th international conference on soil mechanics and foundation engineering. Mexico City: State of the Art Report; 1969. p. 225–290
2. Attewell PB, Woodman JP. Predicting the dynamics of ground settlement and its derivatives caused by tunneling in soil. Gr Eng. 1982; 15(8):13–20, 36
3. Liu J, Hou X. Shield tunneling method. Beijing: China Railway Publishing House;1991
4. Sagaseta C. Analysis of undrained soil deformation due to ground loss. Geotechnique. 1987;37(3):301–20
5. Sagaseta C. Author's reply to Schmidt. Geotechnique. 1988;38(4):647–9
6. Verruijt A, Booker JR. Surface settlements due to deformation of a tunnel in an elastic half plane. Geotechnique. 1996;46(4):753–6
7. Loganathan N, Poulos HG. Analytical prediction for tunneling-induced ground movement in clays. J Geotech Geoenviron Eng. 1998;124(9):846–56
8. Chen F, Hu Z. Analytical prediction of tunneling induced surface movements due to shield-deviation in undrained soil. Rock and Soil Mech. 2004; 25(9):1427–1431
9. Jiang X, Zhao Z. Application of image method in calculating tunneling-induced soil displacement. J Harbin Inst Technol. 2005; 37(6):801–803
10. Wei G. Theoretical study on properties of soil and structure during pipe jacking construction. Hangzhou: Zhejiang University;2005
11. Tang X, Zhu J, Liu W et al. Research on soil deformation during shield construction process. Chin J Rock Mech Eng. 2010; 29(2):417–422
12. Yan B, Yang G, Lin H. Research on shield tunneling parameters optimization and surface subsidence control. Chin J Undergr Space Eng. 2011; 7(S2):1683–1687
13. Wang H, Fu D. Theoretical and test studies on balance control of EPB shields. China Civ Eng J. 2007; 40(5):61–68
14. Jiang X, Zhao S. Accident analysis of prestressed reinforced concrete structures during construction. J Civ Eng Manag. 2005; 22(2):1–4, 12
15. Jia N, Yun T. Elastic mechanics. Guangzhou: South China University of Technology Press;1990
16. Zhao Z. Study of image theory and application of regression method on tunnelling induced soil displacements and stresses. Tianjin: Tianjin University;2004

17. Lee KM, Rowe RK, Lo KY. Subsidence owing to tunneling. I. Estimating the gap parameter. Can Geotech J. 1992; 29:929–940
18. Park KH. Elastic solution for tunneling-induced ground movements in clays. Int J Geomech. 2004;4(4):310–8
19. Qi J. Study on environmental effects and structural performance of shield tunnel. Hangzhou: Zhejiang University;2007
20. Zhu C, Li N, Liu H et al. Analysis of ground settlement induced by workmanship of shield tunnelling. Rock Soil Mech. 2011; 32(1):158–164
21. Guo R, Fang Y, He C. Study on the correlation between stress release and displacement release during tunnel excavation. J Railw Eng Soc. 2010; 144(9):46–50
22. Li Z, Liao S, Dai Z. Theoretical study on synchronous grouting filling patterns and pressure distribution of EPB shield tunnels. Chin J Geotech Eng. 2010; 32(11):1752–1757
23. Bezuijen A, Talmon AM, Kaalberg FJ, et al. Field measurements of grout pressures during tunnelling of the Sophia rail tunnel. Soils Found. 2004;44(1):39–48
24. Li M. Study on the performance of synchronous grouting with new hard slurry and its application in shield construction. SHANG Hai: Tongji University;2002
25. Wei G, Zhou Y, Wei X. Research of influence of EBP shield tunneling parameters on ground uplift. Chin J Undergr Space Eng. 2012; 8(S2):1703–1709
26. Wei G, Xu R. Prediction of longitudinal ground deformation due to tunnel construction with shield in soft soil. Chin J Geotech Eng. 2005; 27(9):1077–1081

Chapter 4
A Theoretical Study on the Influence of Longitudinal Tunneling of Shield on Adjacent Shallow Foundation Buildings

4.1 Introduction

The subway shield tunneling is actually a dynamic process, and the tunnel excavation should result in three-dimensional settlement trough, which means that before the tunnel excavation surface reaches the building foundation, the surface deformation caused by shield tunneling has had an impact on the building, which is easy to be ignored in the process of construction, design and research. At present, the research on the impact of shield construction surrounding environment is focused on the issue of horizontal sedimentation tank in the ground, and most of the studies are not considering the conditions of building existence, studies of the impact on the internal forces of the adjacent buildings are much rarer, which cannot be fully and accurately evaluate the impact of shield tunneling on ground buildings, which is difficult to ensure the safety of ground buildings.

Engineering practice also shows that the deformation of soil and adjacent buildings in the tunneling area is not a process independent of time, but a process changing with time [1, 2]. In the tunneling area, even if it is a certain distance from the building, each step of excavation will affect the longitudinal deformation of the overlying soil, and each longitudinal deformation movement of the soil will affect the external force environment of the building foundation, so the whole system is in motion all the time. The characteristics of the interaction between soil and foundation are also changing. Through constant adjustment of internal force and deformation, the two seek for the equilibrium state at the corresponding moment in each stage of tunneling, as shown in Fig. 4.1.

At present, there is almost no theoretical research on settlement and internal force change of neighboring buildings caused by shield tunneling at home and abroad. This chapter first introduces synergy model research of shield tunnel longitudinal tunneling on the adjacent shallow foundation structures [3, 4], the influence of the shield construction of soil loss as a major cause of land subsidence of shallow foundation buildings is deduced. The mechanical model is derived and solved by the

Z. Ding et al., *Influence of Shield Tunneling on Adjacent Structures and Control Technology*, https://doi.org/10.1007/978-981-19-1134-7_4

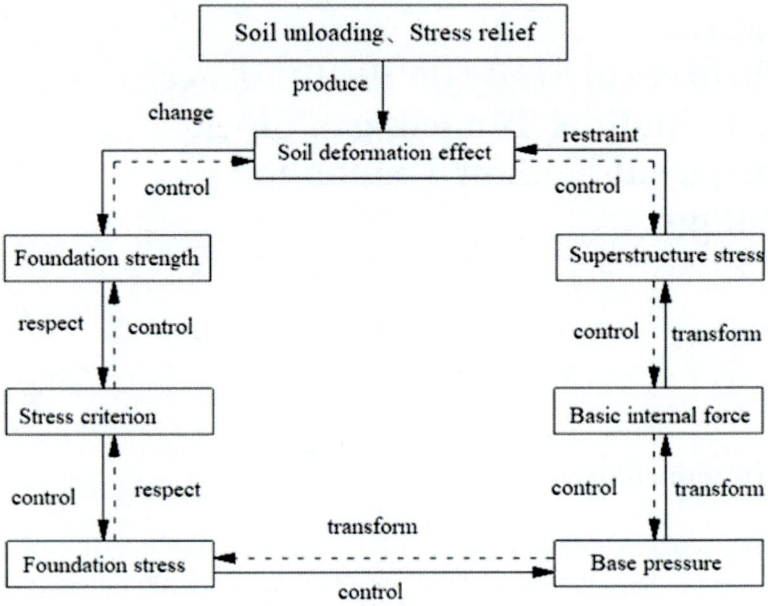

Fig. 4.1 Interaction system between soil deformation and foundation dynamics

numeriacl analysis application software 1stOpt, and the variation rules of longitudinal deformation and internal force are established.

4.2 Theoretical Research on Beam Synergic Action Model Based on Elastic Foundation

4.2.1 The Establishment of the Calculation Model

According to the Ref. [3], the two coordinate systems as shown in Fig. 4.2 are established to simplify the shallow foundation building into a beam on the elastic foundation. The coordinate system of surface deformation is $w_1(j)$-o_1-j. The origin point o_1 is built on the surface directly above the excavation face of the shield tunneling machine. The *X*-axis *j* points to the same direction as the tunneling direction of the shield machine, the ordinate $w_1(j)$ is the surface displacement at point *j*. Displacement coordinate system of the building for $w(x)$-o-x, the origin is based on the left end face of the building, the distance from the o_1 to *j* (Before the excavation face reaches the building, *j* is positive; after the excavation face passes through the building and leaves, *j* is negative), the abscissa of the *x* axis direction and *j* axis pointing to the consistent, deformation of the buildings of any point *x* for $w(x)$, the corresponding point of the surface deformation is $w_1\ (j + x)$.

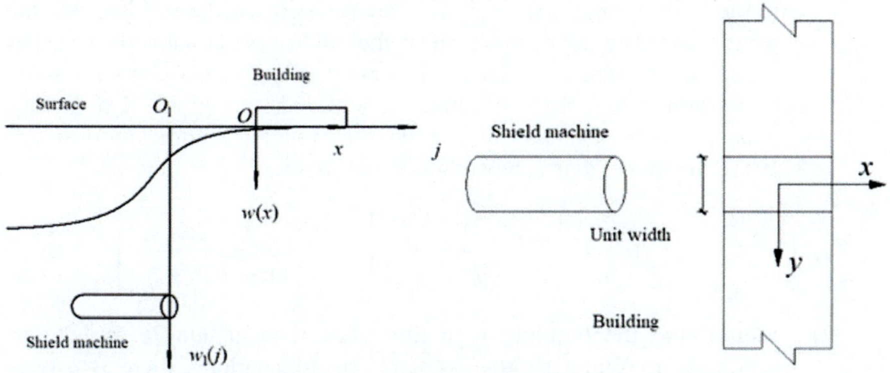

Fig. 4.2 Building and surface settlement coordinate system

(1) Basic assumptions

1) As shown in Fig. 4.3, ground deformation caused by soil loss during shield construction can be calculated by the vertical displacement formula derived in the previous chapter:

$$\begin{aligned} S_{z0} = & \frac{V_{loss1}}{2\pi}\frac{h}{y^2+h^2}\left[1-\frac{x}{\sqrt{x^2+y^2+h^2}}\right] \\ & + \frac{V_{loss2}}{2\pi}\frac{h}{y^2+h^2}\left[1-\frac{x+L}{\sqrt{(x+L)^2+y^2+h^2}}\right] \\ & - \frac{V_{loss3}}{2\pi}\frac{h}{y^2+h^2}\left[1-\frac{x+L}{\sqrt{(x+L)^2+y^2+h^2}}\right] \end{aligned} \tag{4.1}$$

Fig. 4.3 Schematic diagram of soil loss

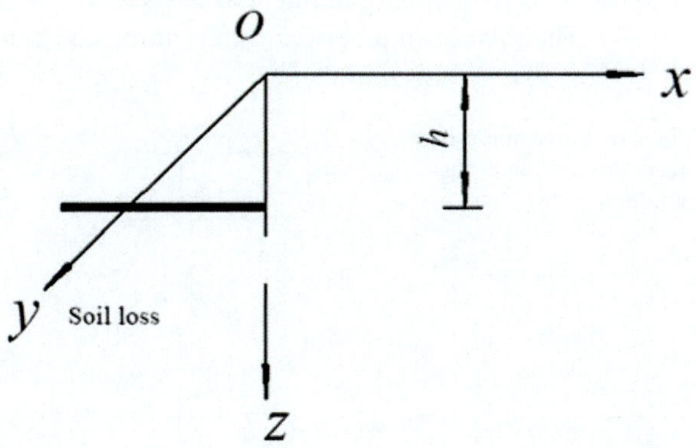

In the formula, V_{loss1}, V_{loss2}, V_{loss3} are respectively the soil loss per unit length caused by excavation surface, the soil loss per unit length caused by shield tail clearance and the soil loss per unit length caused by grouting.

Assuming that the excavation is balanced during shield tunneling, when y = 0, the calculation formula of ground longitudinal deformation above the tunnel axis can be obtained as follows

$$w_1(x) = \frac{V_{loss2} - V_{loss3}}{2\pi h} \times \left[1 - \frac{x + L}{\sqrt{(x + L)^2 + h^2}}\right] \tag{4.2}$$

2) Simplifying the building to a finite beam on an elastic foundation, according to Winkler's theory of elastic foundations, there is a positive correlation between ground reaction σ_dD and the value of the shear foundation, and the calculation formula is as follows

$$\sigma_d(x) = k[w(x) - w_1(j + x)] \tag{4.3}$$

In the formula, k is the subgrade bed coefficient in units of kN/m^3, $\sigma_d(x)$ is the ground reaction at any point at the bottom of a building in units of kN/m^2.

(2) Basic differential equation

As shown in Fig. 4.4, the differential equation of building bending moment is as follows

$$EJ\frac{\mathrm{d}^2 w(x)}{\mathrm{d}x^2} = M(x) \tag{4.4}$$

In the formula, E is the elastic modulus of the building; J is moment of inertia; M (x) is the bending moment at any section in the building.

The relationship between shear force and bending moment is as follows

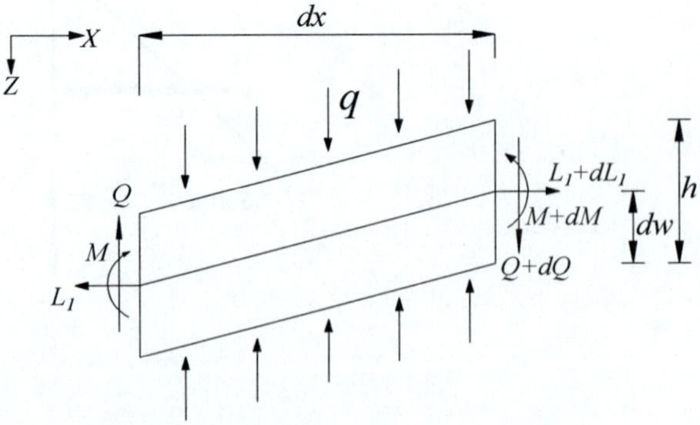

Fig. 4.4 Force analysis of the micro-section of the building

$$Q(x) = \frac{dM(x)}{dx} \tag{4.5}$$

When the transition from a certain section of the building to a section with a distance dx, the increase in shear force is $dQ(x) = (q - \sigma_d)dx$, that is

$$\frac{dQ(x)}{dx} = q - \sigma_d = q - k[w(x) - w_1(j + x)] \tag{4.6}$$

In the formula, q is the vertical load of the building acting on the foundation.

Taking the above equations into consideration, the differential equation of building bending on elastic foundation is obtained as follows

$$EJ\frac{d^4w(x)}{dx^4} = q - k[w(x) - w_1(j + x)] \tag{4.7}$$

Substitute Eq. (4.2) into Eq. (4.7) and get

$$EJ\frac{d^4w(x)}{dx^4} + kw(x) = q + \frac{k(V_{loss2} - V_{loss3})}{2\pi h}\left[1 - \frac{j + x + L}{\sqrt{(j + x + L)^2 + h^2}}\right] \tag{4.8}$$

Equation (4.8) is a simplified theoretical differential equation for the synergistic effect of building foundation, shallow foundation and structure.

(3) The boundary conditions

1) Both ends of the building can be considered as free ends, so the shear force is zero, that is

   If $x = 0$ or $x = l$, then $EJ\frac{d^3w(x)}{dx^3} = 0$. In this formula, l is the length of the building, and the unit is m.

2) Both ends of the building can be considered as free ends, then the bending moment is zero, that is

   If $x = 0$ or $x = l$, then $EJ\frac{d^2w(x)}{dx^2} = 0$.

(4) Solutions to differential equations

1) According to the solution method of homogeneous differential equation, the general solution of homogeneous differential equation $EJ\frac{d^4w(x)}{dx^4} + kw(x) = 0$ is as follows

$$w(x) = e^{-ax}(C_1 \sin ax + C_2 \cos ax) + e^{ax}(C_3 \sin ax + C_4 \cos ax)$$

   In the formula, $a = \sqrt[4]{\frac{k}{4EJ}}$; *C1*, *C2*, *C3* and *C4* are undetermined constants.

2) And then find a particular solution w* to the inhomogeneous differential Eq. (4.8), then the solution of the differential Eq. (4.8) is the general solution plus the particular solution. Since the Eq. (4.8) cannot derive the theoretical analytical solution, 1stOpt software can be used for numerical solution.

4.2.2 Analysis of Theoretical Results

Basic conditions for calculation and analysis: bending stiffness EJ of foundation beam is 1280 MN·m^2, length l of building wall is 20 m, vertical load q of building acting on foundation is 200 kN/m^2, coefficient of soft soil foundation bed is 15000 kN/m^3, diameter of shield machine is 6.34 m, length is 6 m, the inner and outer diameters of the lining are 5.50 and 6.20 m respectively, the width is 1.0 m, the grouting rate is 200%, the grouting loss is 65%, and the tunnel buries depth of tunnel axis h is 9.1 m. Parameters are substituted into the differential equation of the synergetic action buidling foundation, shallow foundation and structure, and the numerical calculation is carried out through 1stOpt software to obtain the building subsidence curve, slope curve, bending moment and shear force. This book selects 6 situations with obvious laws for comparison. The distance between excavation face and the left end of the building is 50m, 10m, 0m, −10m, −20m, −30m, −50 m, as is shown in Figs. 4.5 and 4.6.

It can be concluded from Fig. 4.5a and b that when the building is not affected by shield tunneling, the whole building has been cut into the foundation to a greater depth, which is related to the influencing factors such as foundation soil bed coefficient and vertical loads of the building, which is because the building needs ground reaction to balance the load on the whole building. When the building is affected by shield tunneling, the subsidence deformation is basically the same as the surface deformation at the corresponding position. As the excavation face of the shield is closer to the building, the subsidence of the building gradually increases. When the excavation face of the shield is at the left end of the building, the settlement of the building gradually increases; When the excavation reached about 1/5 position ($j = -4$ m), the subsidence at the left end of the building increased sharply, and the right end began to settle slowly. Due to the large difference in deformation between the left and right sides of the building, the left end of the building inclines the most at this time; as the excavation face of shield tunneling reaches the middle of the building ($j = -10$ m), the whole building has a large incline; When the excavation face is located at the right end of the building ($j = -20$ m), the deformation of the left end of the building tends to be stable, while the settlement of the right end of the building continues. At this time, the tilt of the right end of the building is the largest, which is consistent with the measured settlement and inclination laws in Chap. 2. The settlement of the building is inclined to one side as a whole, and the variation trend of

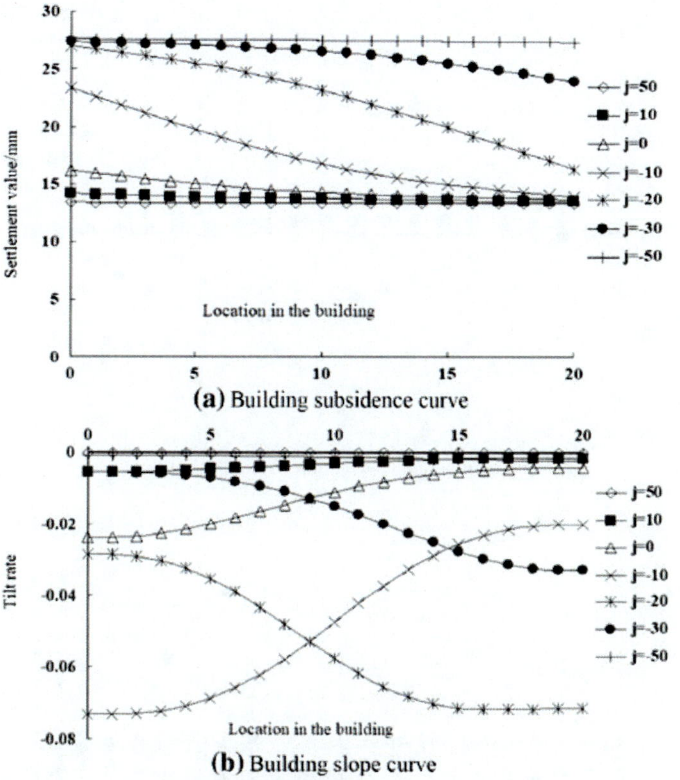

Fig. 4.5 Deformation diagram of the building at different distances from the shield tunneling machine to the building

the subsidence value from the left end to the right end is relatively consistent. This indicates that under this working condition, the shallow foundation building will not be affected by land settlement, resulting in excessive local deformation and failure of the structure.

The additional bending moment and shear force of any section of the building vary greatly during the process of shield tunneling, as shown in Fig. 4.6a and b. When the building deformation occurs, the additional bending moment and shear force in the building body start to produce, when the excavation surface reaches about 1/2 of the building, the position at the right end (the building is located at the maximum negative and positive curvature of the surface subsidence curve), the bending moment value in the middle of the building reaches the maximum, and the distribution of the bending moment presents an approximately symmetric concave or convex curve; when the excavation face is located between the two, the positive and negative bending moment value of the building decreases obviously. Finally, when the roadheader is far away from the building, the settlement of the building tends to be stable, and the value of its additional bending moment also tends to be

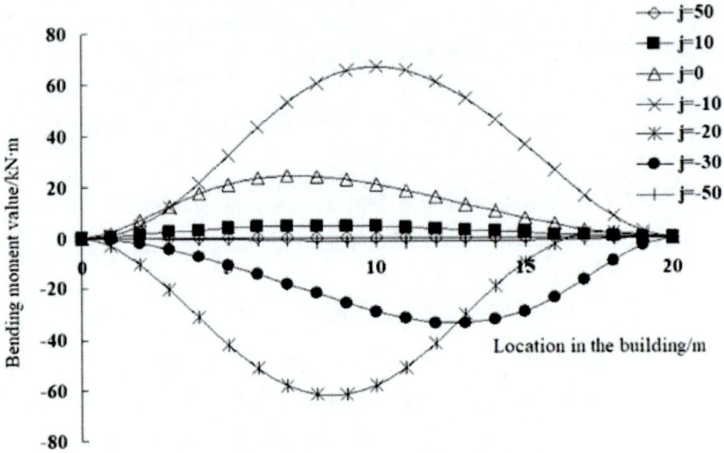

(a) Bending moment distribution within the building

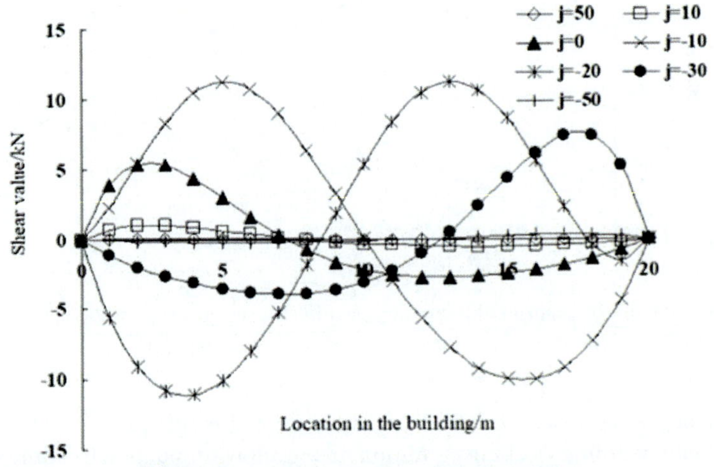

(b) Distribution of shear forces within the building

Fig. 4.6 Additional stress distribution of the building at different distances from the shield machine to the building

zero. From the Formula (4.4), it can be seen that the additional bending moment of the building is determined by the curvature of the building's subsidence curve, while the surface subsidence curve determines the bending degree of the building, which is related to the curvature of the surface subsidence curve. No matter where the excavation face is located, there is always positive and negative shear force in the building, which is also the reason why the length of the subsidence half basin is small while the length of the building is large; when the excavation face reaches about 1/2 of the building, the maximum positive shear force is generated at 1/4 of the building, the maximum negative shear force is generated at 3/4 of the building, and the shear force in the middle is zero. The distribution of shear force is approximately antisymmetric; when the excavation face reaches the right end of the building, it is the opposite. This suggests that measures to strengthen the stiffness in 1/2, 1/4, and 3/4 of the building can be taken to protect the building.

4.2.3 Influence of Various Factors on Additional Stress of Buildings

The additional bending moment and shear force distribution of the building are different when the excavation face of shield tunneling is in different positions. The book considers a special case when the excavation face is at the 1/2 position of the building, that is, $j = -10$ m. At this time, the maximum positive bending moment and maximum shear force are encountered in the building. In this way, the law of change can be reflected more obviously. The variation of additional stress of the building is related to the building type, material, size, foundation property, shield construction technology and other factors. The influences of the flexural stiffness of the building, subgrade bed coefficient, grouting rate and grouting loss rate on internal force are analyzed below.

(1) Flexural stiffness of the building

Different types, materials and sizes of buildings make the different flexural stiffnesses of buildings. With other calculation conditions unchanged, the flexural stiffness of the building is taken as 750, 1000, 1280 and 1500 MN·m^2, respectively. The bending moment and shear force diagram of the building are calculated by 1stOpt software, as shown in Fig. 4.7a and b. It can be seen from Fig. 4.7a that the greater the flexural rigidity of the building, the greater the additional bending moment. This is because the greater the flexural rigidity is, the more difficult the building will be to deform with the surface deformation. The greater the accumulated energy in the building is, the greater the additional bending moment will be. In addition, with the increase of the flexural rigidity of the building, the increase of additional bending moment becomes slow. It can be seen from Fig. 4.7b that the change rule of shear force is similar to the bending moment. It can be seen that increasing the stiffness of the building is effective in resisting the additional deformation caused by the influence of shield tunneling, but there is a certain effective influence range.

(2) Subgrade bed coefficient

In this book, subgrade bed coefficient k in soft soil area is respectively 5000 kN/m^3, 10,000 kN/m^3, 15,000, and 20,000 kN/m^3. Other calculation conditions remain unchanged, and the relation curve between of the building bending moment and

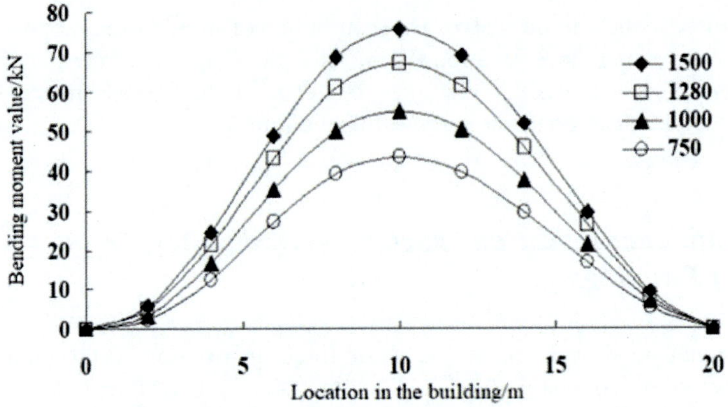

(a) Bending moment distribution curve of foundation beam with different bending stiffnesses

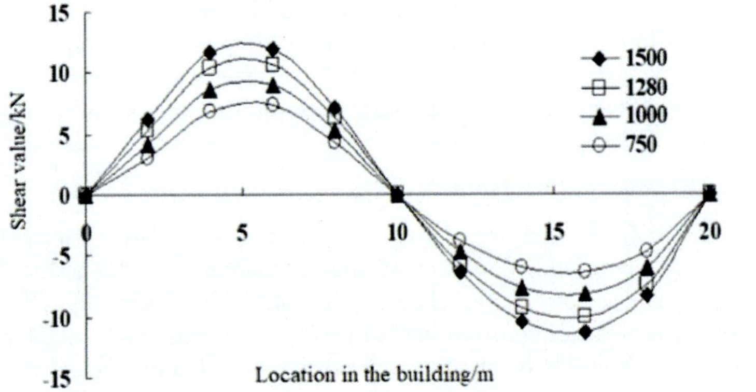

(b) Shear distribution curve of foundation beam with different bending stiffnesses

Fig. 4.7 Internal force distribution of buildings with different flexural stiffnesses

shear distribution curve and subgrade bed coefficient can be obtained, as shown in Fig. 4.8a and b. It can be seen from the figure that the softer the foundation, the smaller the subgrade bed coefficient, the smaller the additional bending moment and shear force of the building, but the larger the reduction. Because the foundation coefficient is small, the easier the building is to cut into the foundation, the more even the distribution of foundation reaction, which reduces the additional stress inside the building. This indicates that under the same surface subsidence condition, the buildings in sandy soil are more likely to be damaged than those in soft clay. At the same time, grouting should be carried out on the soft soil foundation to reduce the variation of additional stress of the building. It is necessary to reinforce the soft soil foundation evenly. Otherwise, the effect may not be as good as that without reinforcement.

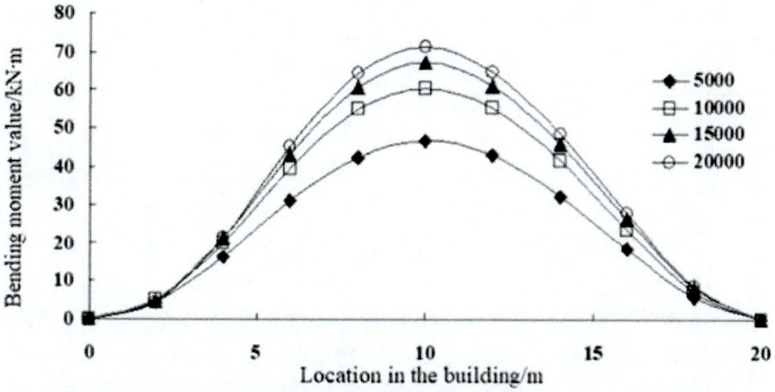

(a) Bending moment distribution curves of foundation beams with different subgrade bed coefficients

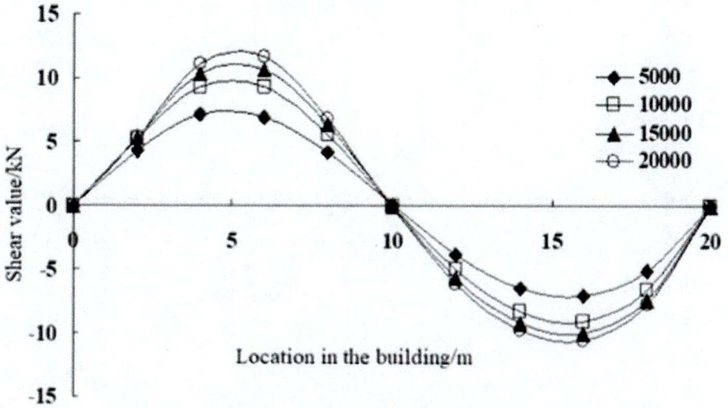

(b) Shear distribution curves of foundation beams with different subgrade bed coefficients

Fig. 4.8 Internal force distribution of buildings with different subgrade bed coefficients

(3) Grouting rate and grouting loss

The book analyzes the grouting rate and grouting loss in the construction process and finds that the construction process has a greater impact on the internal force and deformation of the building. It can be seen from Figs. 4.9a and b, 4.10a and b that the larger the grouting rate is, the smaller the maximum bending moment and shear force of the building will be. This is because with the increase of the grouting rate, the soil loss will be reduced, thus the ground settlement will be reduced and the additional internal force of the building will also be reduced. The larger the grouting loss is, the larger the maximum bending moment and shear force of the building are, which indicates that the grouting rate and grouting loss must be controlled well in the process of shield tunneling.

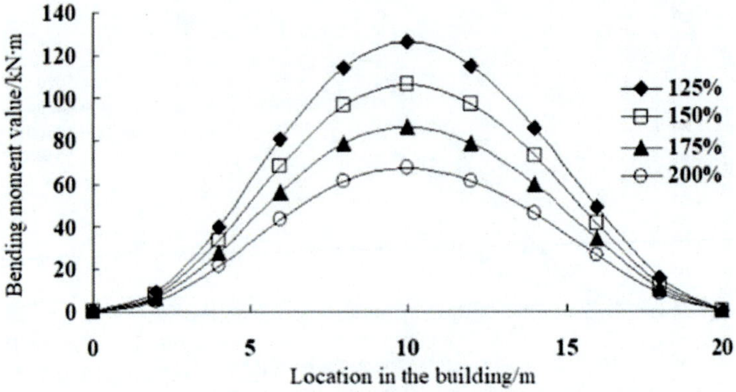

(a) Bending moment distribution curve

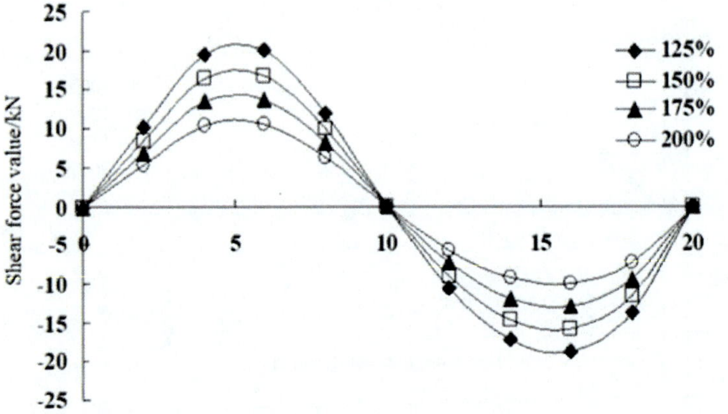

(b) Shear distribution curve

Fig. 4.9 Internal force distribution of buildings with different grouting rates

According to the comparative analysis of Figs. 4.7a, 4.8a, 4.9a and 4.10a, it is found that the influence of construction technology is greater than that of building and foundation reinforcement on the internal force change of the building, which puts forward a more meaningful idea for the protection of the building: Firstly, the ground settlement is reduced from the construction technology, and then the building foundation is strengthened uniformly, and then the building body is strengthened if the conditions permit. During the process of excavation, it is necessary to strengthen the observation of the settlement and inclination of the building, especially when the excavation face is at the position of 1/2 of the building.

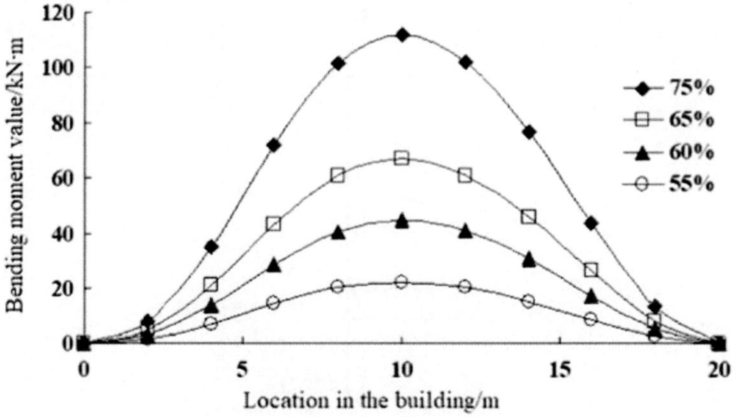

(a) Bending moment distribution curve

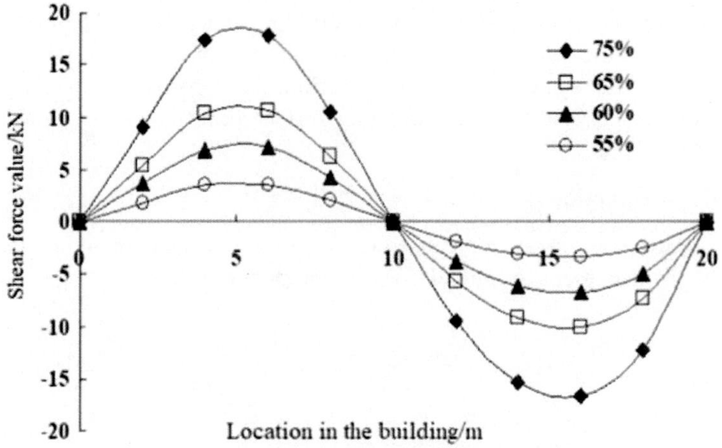

(b) Shear force distribution curve

Fig. 4.10 Internal force distribution of buildings with different grouting loss rates

4.3 Theoretical Research based on the Synergetic Action Model of Shear and Bending Beam of Elastic Foundation

4.3.1 Establishment of Calculation Model

The double coordinate system in the previous section is still adopted, and the frame structure with strip foundation is considered as the building. The surface deformation coordinate system is w_1 (j) -o_1-j, and the origin point o_1 is built on the surface directly above the excavation face. The x-axis j points to the same direction as the advance

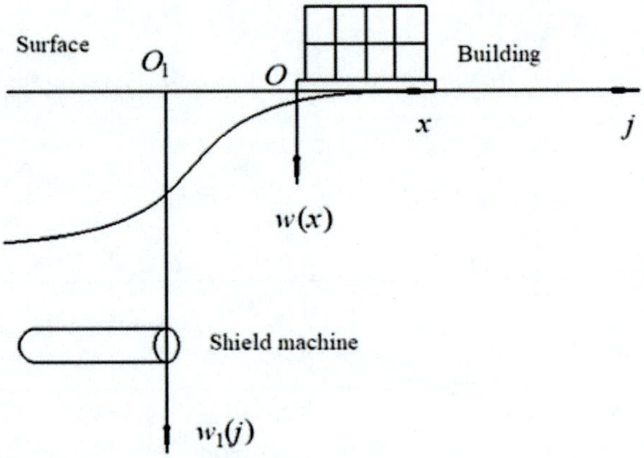

Fig. 4.11 Building and surface settlement coordinate system

direction of the shield, and the *y*-axis w_1 (*j*) is the surface displacement of point *j*; the deformation coordinate system of the building is *w*(*x*)-*o*-*x*. The origin is built on the ground at the left end of the building, and the distance from o_1 is *j*. The *x*-axis direction is the same as the *j*-axis direction, the deformation of any point *x* in the building is *w(x)*, and the corresponding surface point is $w_1(j + x)$, as shown in Fig. 4.11.

(1) Basic assumptions

1) Ground deformation caused by soil loss is still calculated by the vertical displacement formula derived in the previous chapter:

$$w_1(x) = \frac{V_{loss2} - V_{loss3}}{2\pi h} \times \left[1 - \frac{x + L}{\sqrt{(x + L)^2 + h^2}}\right] \tag{4.9}$$

In the formula, V_{loss2} and V_{loss3} are respectively the soil loss per unit length caused by shield tail clearance and the soil loss per unit length caused by grouting.

2) The building is still regarded as the beam on the elastic foundation, and the strip foundation frame structure above the axis of the shield tunnel is simplified, as shown in Fig. 4.12.

Building foundation reaction σ_d is proportional to the value of the shear in foundation:

$$\sigma_d(x) = k[w(x) - w_1(j + x)] \tag{4.10}$$

In the formula, *k* is the subgrade bed coefficient (kN/m^3), *d(x)* is the foundation reaction (kN/m^2) at any point at the bottom of the building.

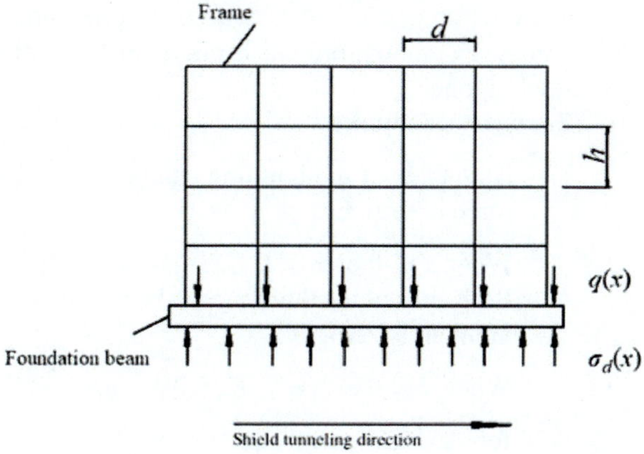

Fig. 4.12 Simplified view of frame structure in tunneling area

(2) Fundamental differential equation

According to the Ref. [4], the strip foundation building is simplified as a shear beam on elastic foundation constrained by superstructure, the differential equation of the shear beam is as follows

$$EJ\frac{\mathrm{d}^4w(x)}{\mathrm{d}x^4} - (GF + g)\frac{\mathrm{d}^2w(x)}{\mathrm{d}x^2} = q(x) - \sigma_d(x) \tag{4.11}$$

In the formula, *EJ* is the bending stiffness of the foundation beam (kN·m^2), *GF* is the vertical shear stiffness of the frame structure (kN), *G* is the constraint line stiffness of the bottom column end (kN), and *q(x)* is the vertical load acting on the foundation by the building (kN/m^2).

In addition, $GF = \frac{12}{d\left(\frac{1}{K_b}+\frac{1}{K_c}\right)}$, $K_b = \frac{\Sigma EI_b}{d}$, $K_c = \frac{\Sigma EI_c}{h}$, $g = \frac{6K_{c1}}{d}$.

In the formula, $\sum EI_b$ is the sum of bending stiffness of each floor beam in the same opening (kN·m^2); $\sum EI_c$ is sum of bending stiffness of each layer of the same column (kN·m^2); K_{c1} is the bottom column stiffness of the frame structure (kN·m); *d* is column spacing (m), *h* is floor height (m).

Equations (4.9) and (4.10) can be substituted into Eq. (4.11) to obtain the flexural differential equation of shear beam on elastic foundation

$$\begin{aligned} &EJ\frac{\mathrm{d}^4w(x)}{\mathrm{d}x^4} - (GF + g)\frac{\mathrm{d}^2w(x)}{\mathrm{d}x^2} + kw(x) \\ &= q + \frac{k(V_{loss2} - V_{loss3})}{2\pi h}\left[1 - \frac{j + x + L}{\sqrt{(j + x + L)^2 + h^2}}\right] \end{aligned} \tag{4.12}$$

Equation (4.12) is the theoretical differential equation of building foundation, strip foundation and frame structure acting together above the axis of shield tunnel.

(3) Boundary conditions

1) Both ends of the building can be considered as free ends, then the shear force is zero, i.e.

When $x = 0$ or $x = l$, $EJ\frac{d^3w(x)}{dx^3} = 0$, l is the length of the building (m).

2) Both ends of the building can be considered as free ends, then the bending moment is zero, i.e.

When $x = 0$ or $x = l$, $EJ\frac{d^2w(x)}{dx^2} = 0$.

(4) Solutions to differential equations

(1) Find the general solution of the homogeneous differential equation.

According to the solution of Euler's equation, let $w = e^{\lambda x}$, $b = -\frac{GF+g}{EJ}$, $q = \frac{k}{EJ}$, The λ is the undetermined coefficient, giving $\lambda^4 + b\lambda^2 + q = 0$, the resulting solution is $\lambda_{1,2} = \pm\frac{1}{2}\sqrt{-2b + 2\sqrt{(b^2 - 4q)}}$, $\lambda_{3,4} = \pm\frac{1}{2}\sqrt{-2b - 2\sqrt{(b^2 - 4q)}}$, so the homogeneous solution is going to be:

$$w(x) = c_1 e^{d1x} + c_2 e^{-d1x} + c_3 e^{d2x} + c_4 e^{-d2x}$$

In the formula: $d1 = \pm\frac{1}{2}\sqrt{-2b + 2\sqrt{(b^2 - 4q)}}$, $d2 = \pm\frac{1}{2}\sqrt{-2b - 2\sqrt{(b^2 - 4q)}}$; c_1, c_2, c_3 and c_4 are undetermined coefficient.

(2) Then find a particular solution of the inhomogeneous differential Eq. (4.12), then the solution of the differential Eq. (4.12) is the general solution plus the particular solution. Since the Eq. (4.12) cannot derive the theoretical analytical solution, 1stOpt software can be used for numerical solution.

4.3.2 Additional Stress of the Building

(1) The additional bending moment of the strip foundation is

$$M(x) = EJ\frac{d^2w(x)}{dx^2} \tag{4.13}$$

(2) Additional shear force on strip foundation is

$$Q(x) = EJ\frac{\mathrm{d}^3 w(x)}{\mathrm{d}x^3} \tag{4.14}$$

(3) Bending moment of additional constraint line generated by bar foundation constraint at the bottom column end of the frame is

$$m_i = g_i \frac{\mathrm{d}w(x)}{\mathrm{d}x} \tag{4.15}$$

where, m_i is the bending moment (kN) of the constraint line at the bottom column end of the column i of the frame, and g_i is the rigidity (kN) of the constraint line at the bottom column end of the column i.

(4) Additional shear force of frame structure beam

The additional total shear force on the frame beam between two adjacent columns can be expressed as

$$v_j = (GF_j + g_j)\frac{\mathrm{d}w(x)}{\mathrm{d}x} \tag{4.16}$$

where, v_j is the additional total shear force (kN) of the beam in the opening j, and GF_j is the total shear stiffness (kN) of the beam in the opening j.

(5) Internal forces of the beam and column of the frame structure

The vertical displacement S_i and angular displacement θ_i at the junction of foundation beam and column caused by surface movement can be solved by using the approximate method to solve the internal force of beam and column of the frame structure [4], i.e.

1) Shear force of beam and column

The additional total shear force on the frame can be calculated through Eq. (4.16), and then distributed to the beams on each floor according to the shear stiffness of the beams on each floor between the two columns, and the shear force in the span of the frame beam is

$$v_{jk} = v_j \times \frac{GF_{jk}}{GF_j} \tag{4.17}$$

where, v_{jk} is the shear force in the span of the frame beam on the k floor in the opening j, and GF_{jk} is the shear stiffness (kN) of the frame beam on the k floor in the opening j.

Among them

$$GF_j = \frac{12}{d_j\left(\frac{1}{K_{bj}} + \frac{1}{K_{ci}}\right)},\ K_{bj} = \frac{\Sigma EI_{bj}}{d_j},\ K_{ci} = \frac{\Sigma EI_{ci}}{h_i},$$

$$GF_{jk} = \frac{12}{d_j\left(\frac{1}{K_{bjk}} + \frac{1}{K_{cik}}\right)},\ K_{bjk} = \frac{\Sigma EI_{bjk}}{d_j},\ K_{cik} = \frac{\Sigma EI_{cik}}{h_i},$$

where, $\sum EI_{bj}$ is the sum of the bending stiffness of the frame beam of each floor in the opening j (kN·m^2); $\sum EI_{ci}$ is the sum of the bending stiffness of the frame column of each floor in the opening i (kN·m^2); EI_{bjk} is the bending stiffness of the frame beam of the floor k in the opening j (kN·m^2); EI_{cik} is the bending stiffness of the frame column of the floor k in the opening i (kN·m^2).

The end i of the column at the bottom of the frame will generate constraints on the strip foundation, and the equivalent shear force generated by the constraint can be expressed as $v_{Ii} = g_i\theta_i$.

2) The bending moment of beam and column of the frame of each layer can be solved according to the method of reverse bending point

The bending moment of the beam end can be expressed as:

$$M_{jk} = v_{jk} \times d_j/2 \tag{4.18}$$

The end of the bottom column imposes constraints on the strip foundation, and the constraint bending moment of the foot of the column generated by the constraint is expressed as:

$$M_{1i} = g_i\theta_i \times d_i/2 \tag{4.19}$$

4.3.3 Calculation and Analysis of Examples

The basic conditions of calculation: the building is reinforced concrete frame structure, using C30 concrete pouring. There are 4 floors above ground, each height is 3.6 m (including plate thickness), and the spacing is 4 m along the direction of shield tunneling. The size of column is 300 × 300 mm, the size of beam is 300 × 550 mm and the thickness of floor is 100 mm. The foundation is strip foundation and the cross section size is 1000 × 800 mm. The elastic modulus of reinforced concrete soil is 30,000 MPa, the vertical load of the building on the foundation is 200 kN/m^2, the coefficient of soft soil foundation bed k is 15,000 kN/m^3, the diameter of shield tunneling machine is 6.34 m, the length is 6 m, the inner and outer diameters of the lining are 5.50 and 6.20 m respectively, the width is 1.0 m, the grouting rate is 200%, the grouting loss is 65%, and the depth of tunnel axis h is 9.1 m.

The transverse branches of the frame structure are named as 1, 2, 3, 4 and 5; Each column is named as I, II, III, IV, V and VI; the vertical floors are named A, B, C and D, as shown in Fig. 4.13.

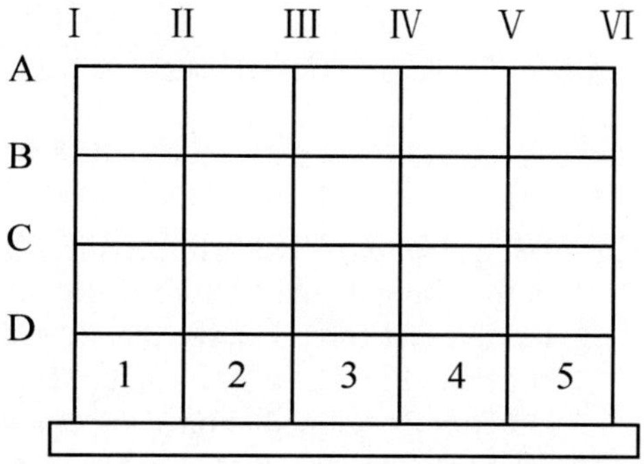

Fig. 4.13 Schematic diagram of a framed building

(1) Calculation of relevant parameters

Column:

$$I_c = \frac{bh^3}{12} = \frac{0.3^4}{12} = 0.000675\,\text{m}^4$$

$$EI_c = 20.25 \times 10^3 \text{kN} \cdot \text{m}^2$$

$$K_c = \frac{\Sigma EI_c}{h} = \frac{4 \times 20.25 \times 10^3}{3.6} = 22.5 \times 10^3\,\text{kN} \cdot \text{m}$$

Beam:

$$I_b = \frac{bh^3}{12} = \frac{0.3 \times 0.55^3}{12} = 0.00416\,\text{m}^4$$

$$EI_b = 124.781 \times 10^3 \text{kN} \cdot \text{m}^2$$

$$K_b = \frac{\Sigma EI_b}{d} = \frac{4 \times 124.78125 \times 10^3}{4} = 124.781 \times 10^3\,\text{kN} \cdot \text{m}$$

$$GF = \frac{12}{d\left(\frac{1}{K_b} + \frac{1}{K_c}\right)} = 57188.096kN$$

$$g = \frac{6K_{c1}}{d} = \frac{6 \times 22.5 \times 10^3}{4} = 33.75 \times 10^3 kN$$

$$GF + g = 90938.096\text{kN}$$

Foundation beam:

$$J = \frac{bh^3}{12} = \frac{1 \times 0.8^3}{12} = 0.0427m^4$$

$$EJ = 1.28 \times 10^6 \text{kN} \cdot \text{m}^2$$

(2) Using 1stOpt software to solve Eq. (4.12) numerically, the subsidence curve of the building, the additional stress distribution curve of strip foundation and the internal force change curve of the beam and column are calculated, as shown in Fig. 4.14a–e. It can be seen from the settlement curve that the frame structure is inclined to one side as a whole, and the settlement at each point is relatively uniform, that is, the shallow foundation building will not be affected by the sharp subsidence of the ground under this working condition, so that the local deformation of the structure is large and the damage is caused. Under the influence of shield tunneling, the deformation of the building is consistent with the surface deformation at its corresponding position. As the excavation moves toward the building, the subsidence of the building increases. The maximum amount of subsidence is achieved when the shield machine passes through the building. When the shield tunneling machine continues, the influence on the subsidence of the frame structure gradually weakens. It can be seen from Fig. 4.14a that when the excavation face of shield is 10 m below the building, the settlement difference between the head and the tail of the building is the largest.

According to the distribution of bending moment and shear force, when the excavation face reaches the position of about 1/2 below the building and the right end, the bending moment received in the middle of the strip foundation reaches the maximum, while the shear force reaches the maximum at the position of about 1/4 and 3/4 of the strip foundation, and there are positive and negative shear forces respectively. The internal force of the building changed little when the shield tunnel is 10m away from the left end of the building, but changed a lot when is 10m away from the right end. It can be seen that more attention should be paid to the internal force change of the building after the shield tunneling through the shallow foundation frame structure, which makes the building more vulnerable to damage.

Figure 4.14d and e show the change curve of mid-span shear force and beam end bending moment of the beam along with shield tunneling. When the excavation face is under the building, the internal force of the frame structure of the building changes greatly. It can be seen that the shear force and bending moment gradually increase with the approach of the excavation face. When the excavation face reaches the vicinity of the right under the beam, the shear force and bending moment reach their maximum value, and then the bending moment and shear force gradually decrease. The whole curve is approximately symmetric. The maximum shear force of the C-storey beam is greater than that of the foundation beam, indicating that the stress variation of the frame beam in the driving zone is greater than that of the shallow foundation beam, which should be paid attention to in the construction and design. The constraint moment law of the bottom column on the foundation reflected in

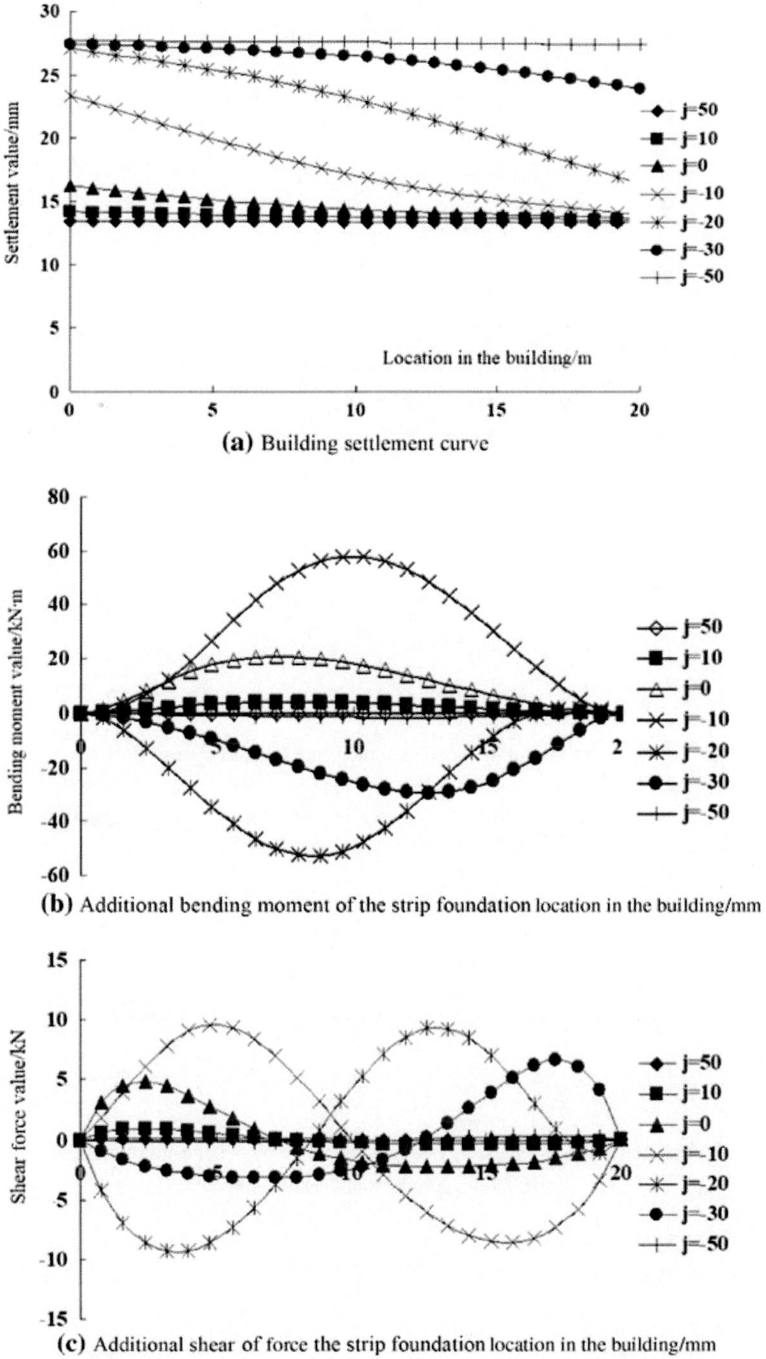

Fig. 4.14 Variation of additional deformation and internal force of frame building

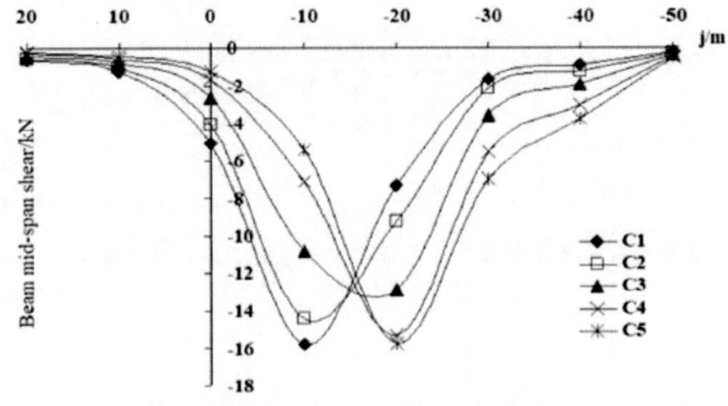

(d) Mid-span shear force of beam in Floor C

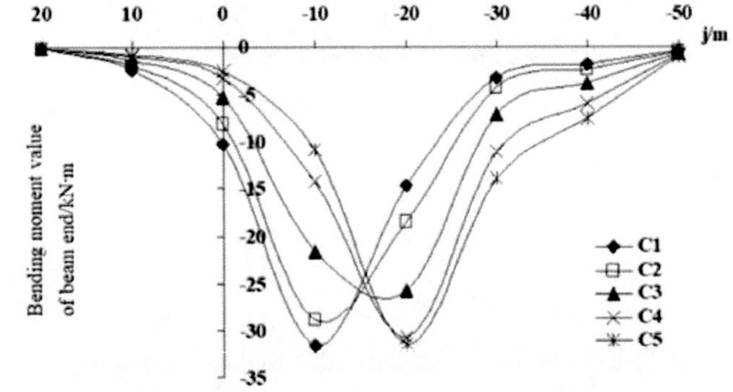

(e) Bending moment at the beam end of Layer C

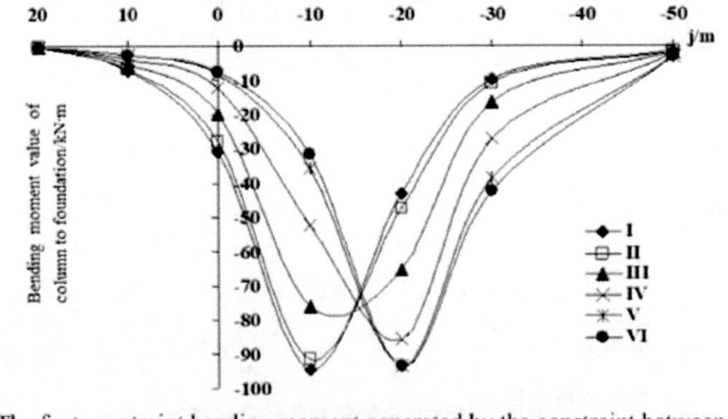

(f) The foot constraint bending moment generated by the constraint between the bottom column and the strip foundation

Fig. 4.14 (continued)

Fig. 4.14f is consistent with the internal force variation diagram of the frame beam, but the maximum bending moment value is larger, indicating that the connection between the column and the foundation is also more vulnerable to damage during tunneling. In addition, from Eqs. (4.15) and (4.16), it can be seen that the internal force of the beam-column of the upper frame structure of the building is mainly determined by the tilt of the building. The greater the tilt rate is, the larger the mid-span shear force of the beam, the bending moment of the beam end and the constrained bending moment of the bottom column on the foundation will naturally be. Therefore, during the construction process, it is necessary to strengthen the monitoring of the settlement difference and tilt rate at the end of the building, so as to better control the influence of shield tunneling on the internal force of the adjacent shallow foundation frame structure.

4.4 Summary

During the construction of shield tunnel in soft soil area, the adjacent buildings will generate large deformation and additional stress, and the deformation and internal force variation are generally related to the type, material, size, foundation property and shield construction technology of the building. In this book, the building is simplified into a beam on elastic foundation, including pure bending beam and shear bending beam. Based on the calculation theory of soil loss, the mechanical theoretical model of the synergistic effect of the foundation, shallow foundation and structure of the building in the area affected by shield tunneling construction is established. The differential equation was solved numerically with 1stopt software, and the settlement change, tilt change and additional stress distribution law of adjacent buildings during the tunneling process of shield tunneling machine were analyzed theoretically.

(1) The law of settlement and deformation of buildings in shield tunneling area is basically consistent with that of surface settlement. When the whole building appears inclined, the trend of subsidence change from left end to right end is relatively consistent, and the building is not prone to local damage. Strengthening the central part of the building can achieve the effect of protection. The maximum bending moment value and shear force value of the building are closely related to the flexural stiffness foundation bed coefficient and grouting rate of the building. The additional stress of the building and the damage of the building can be reduced by softening the foundation, strengthening the building body and improving the construction technology of shield tunnel.

(2) For the frame structure with shallow foundation, the shear force and bending moment gradually increase with the approach of the excavation face. When the excavation face reaches the vicinity of the building directly below, the shear force and bending moment reach the maximum value, and then the bending moment and shear force gradually decrease, and the whole curve is approximately symmetric. The maximum bending moment and shear force of the floor

beam are greater than those of the foundation beam, which indicates that the stress of the frame beam in the driving zone is greater than that of the shallow foundation beam. The constraint moment law of the bottom column on the foundation is consistent with the diagram of the internal force variation of the frame beam, but the maximum moment value is larger, which indicates that the connection between the column and the foundation is more vulnerable to damage during the tunneling.

(3) The internal force of the beam-column of the upper frame structure of the building is mainly determined by the tilt of the building. The greater the tilt rate is, the greater the mid-span shear force of the beam, the bending moment of the beam end and the constrained bending moment of the bottom column on the foundation will be. Therefore, during the construction process, it is necessary to strengthen the monitoring of the settlement difference and tilt rate at the end of the building, and better control the influence of shield tunneling on the internal force of the adjacent shallow foundation frame structure.

The synergistic mechanism of building foundation, foundation and structure is quite complex, the cooperative model of building foundation, foundation and structure proposed in this book is a planar model, and it only considers the working conditions of elastic foundation model and strip foundation frame structure. there's a lot more that needs to be done with the model. Only by truly revealing the law of deformation and internal force change of adjacent buildings under the influence of shield tunneling disturbance, can structures in the tunneling area be designed and constructed in a targeted manner, and for the existing structure, it's necessary to take reasonable reinforcement measures to minimize the disturbance and damage of adjacent structures during construction, safe and reliable protection of the existing buildings.

References

1. Sun Y, Guan F. Shield tunnel construction induced influence on the settlement of masonry buildings. China Rail Sci. 2012;33(4):38–44.
2. Xu Z, Han Q, Zheng G. Field monitoring and analysis of effects of metro tunnels under historic buildings. Chinese J Geotechn Eng. 2013; 35(2):364–74.
3. Tan Z, Deng KZ. Coordinating work model of ground, foundation and structure of building in mining area. J China Univ Min Technol. 2004;33(3):264–7.
4. Xia JW, Yuan YS, Dong ZZ. Mechanism study on subsoil-strap footing-framework interaction in mining subsidence area. Chinese J Geotechn Eng. 2007;29(4):537–41.

Chapter 5
Theoretical Research on the Influence of Longitudinal Tunneling of Shield on Adjacent Buildings with Short Pile Foundation

5.1 Introduction

With the shortage of urban land, underground space has been further developed and utilized, and higher requirements have been put forward for shield construction. At present, there are a large number of buildings built in the 1980 and 1990s in the urban center area, most of which are shallow foundation brick and masonry structures, frame buildings and short pile brick and masonry structures. For example, the tunnel project of Wulin Square Station—Cultural Square Station—Genshanmen Station of Hangzhou Metro Line 1 will pass through several residential communities successively, passing directly below and laterlally below from more than 20 buildings [1]. Wuhan Metro # 2 zhongshan Park Station—Xunlimen Station—Jianghan Road Station tunnel line project also passed under more than 10 buildings [2]. Academician Sun Jun classified the tunneling problem of subway shield tunnels under buildings as the most difficult and outstanding environmental geotechnical problem in the underground engineering activities in soft soil areas [3].

At present, there are many researches on soil displacement caused by shield tunneling, but few theoretical researches on shield tunneling on adjacent pile foundation buildings are found, which mainly focus on numerical simulation analysis without considering the influence of superstructure and its stiffness. In terms of theoretical research, Wei Gang [4] used Mindlin solution to study the distribution law of additional load caused by additional thrust on the front of pipe jacking tunnel, friction between the roadheader and subsequent pipelines and soil on the adjacent pile foundation, and believed that the change law of additional load was closely related to the relative position of pile foundation and roadheader. Wei Xinjiang [5] applied the "sourcing-sink method" theory to infer the calculation formula of three-dimensional additional stress caused by soil loss in double-circular shield tunnel, and studied the additional thrust on the front of the double-circular shield machine, the friction between the shield shell and soil, and the distribution law of additional load caused by soil loss on the adjacent pile foundation. The theoretical research on the influence of shield tunnel construction on adjacent pile foundation buildings is

Z. Ding et al., *Influence of Shield Tunneling on Adjacent Structures and Control Technology*, https://doi.org/10.1007/978-981-19-1134-7_5

still mainly focused on pile foundation, and the research considering the influence of superstructure stiffness has not been reported yet. As the tunneling under shield is usually under shallow foundations or short pile structures, further theoretical analysis and research are needed.

Based on the synergetic model, this chapter takes short piles and soil as springs with different stiffness, simplifies the brick-concrete and frame buildings into pure bending beam and shear bending beam, and deduces the mechanical model and theoretical solution of the synergetic effect of subgrade? foundation and structure of short pile foundation buildings. The influence of longitudinal tunneling of shield tunnel on adjacent short pile foundation buildings is studied, and the numerical analysis software 1stOpt is used to solve the problem, and the longitudinal deformation rule and internal force change rule of the building are analyzed.

5.2 Theoretical Research on Beam Synergy Model Based on Elastic Foundation

The two coordinate systems established in the previous chapter are still adopted. The ground settlement coordinate system is $w_1(j)$-O_1-j, and the settlement coordinate system of the building is $w(x)$-O-x. The x-axis direction is consistent with the j-axis direction. The settlement value of any point x in the building is $w(x)$, and the settlement value of the corresponding surface point is $w_1(j + x)$, as shown in Fig. 5.1.

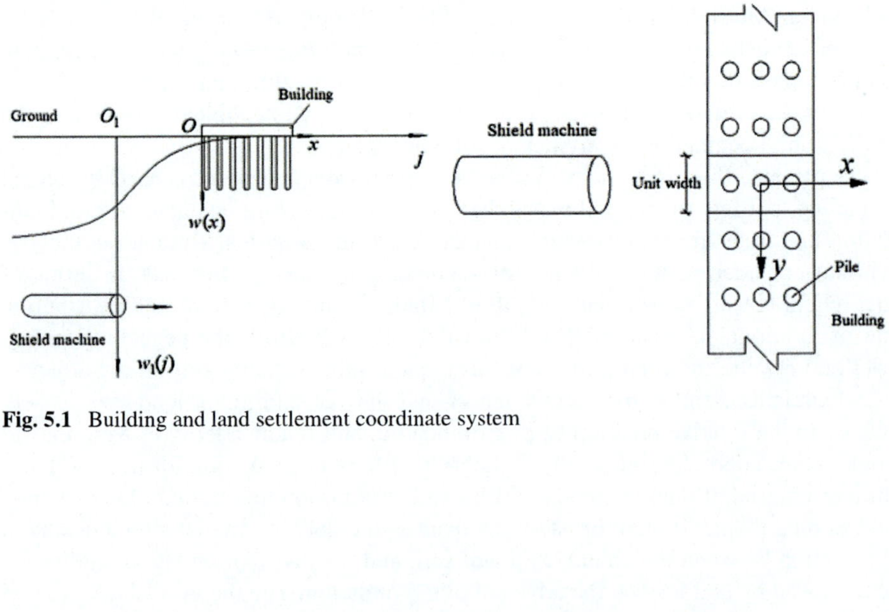

Fig. 5.1 Building and land settlement coordinate system

5.2.1 Establishment of Calculation Model

(1) **Basic assumptions**

1) It is assumed that the excavation rate is balanced during shield construction, and the vertical displacement formula of the ground is the same as Sect. 4.2:

$$w_1(x) = \frac{V_{loss2} - V_{loss3}}{2\pi h} \times \left[1 - \frac{x + L}{\sqrt{(x+L)^2 + h^2}}\right] \tag{5.1}$$

2) According to the theory of Winkle's elastic foundation, the dynamic force 6d of a building is directly proportional to the value of the cut foundation. The formula is as follows

$$\sigma_d(x) = k[w(x) - w_1(j + x)] \tag{5.2}$$

(2) **Basic differential equation**

According to the Ref. [6], for short pile foundation buildings, the plane strain problem is still considered. As shown in Fig. 5.2, buildings and pile foundations with unit length and width of B are simplified as beams on elastic foundations. Soil between piles and short piles can be regarded as linear springs with different stiffness. The spring stiffness coefficient of soil between piles is k_s (kN/m^3), and the spring stiffness coefficient of short piles is K_p (kN/m). The value can be calculated according to the slope of the cutting line on the *Q*-*s* curve of the static load test on the field pile, or according to the relevant norms through reference of experience; Then the average stiffness coefficient of the element where the spring distribution of short pile foundation is $k_p = k_s + \frac{K_p}{dB}$ (kN/m^3), where d is the diameter of short pile foundation.

A differential equation for the bending of a building on an elastic foundation

$$EJ\frac{\mathrm{d}^4 w(x)}{\mathrm{d}x^4} = q - \sigma_d(x) \tag{5.3}$$

Simultaneous Equations (5.1), (5.2) and (5.3), we can get the equation as follows:

$$EJ\frac{\mathrm{d}^4 w(x)}{\mathrm{d}x^4} + k'w(x) = q + \frac{k'(V_{loss2} - V_{loss3})}{2\pi h}\left[1 - \frac{j + x + L}{\sqrt{(j + x + L)^2 + h^2}}\right] \tag{5.4}$$

Formula (5.4) is differential equation of synergistic action of building, foundation and foundation above tunnel. The value of k' in the soil range is k_s and in the pile range is k_s.

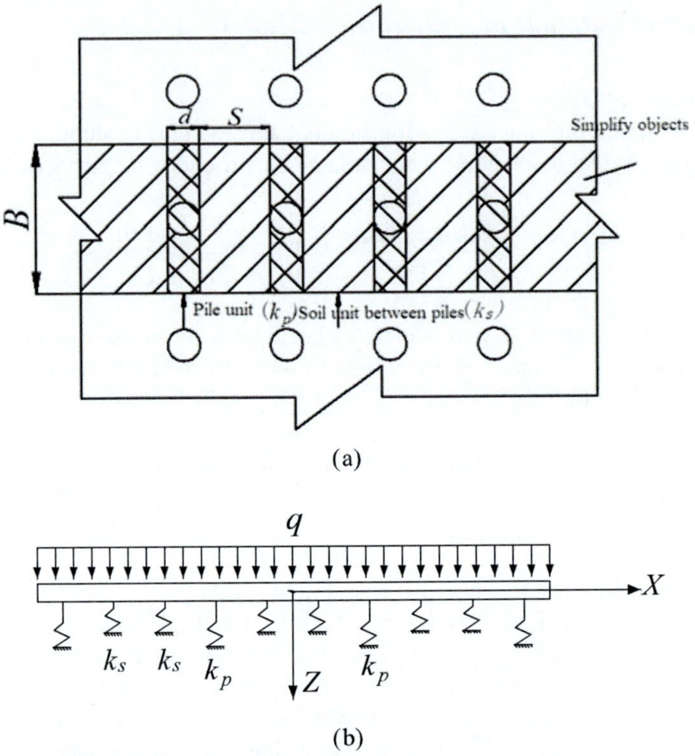

Fig. 5.2 Model diagram of pile foundation building

(3) Solution of differential equation

According to the Ref. [6], according to the solution method of homogeneous differential equation, homogeneous differential equation

$$EJ\frac{d^4 w(x)}{dx^4}+kw(x)=0 \tag{5.5}$$

It can be converted into

$$\frac{d^4 w(x)}{dx^4}+\alpha^4 w(x)=0 \tag{5.6}$$

In the formula : $\alpha=\sqrt[4]{\frac{k}{EJ}}$

For formula (5.6), the power series method can be used to solve. Suppose:

$$w_1=\sum_{n=0}^{\infty} a_n x^n = a_0 + a_1 x + a_2 x^2 + \ldots . a_n x^n + \ldots \ldots \tag{5.7}$$

where: $a_i (i = 0 \sim n)$ is the undetermined coefficient. After 1–4 derivatives are obtained from Eq. (5.7), and then substituted into Eq. (5.6), the comparison coefficient can be obtained as follows:

$$a_{n+4} = a_1^{n+4}\left(\sum_{j=0}^{3}\frac{a_j}{a_1^j}c_{j,n+4}\right)(n \geq 0) \tag{5.8}$$

$$\text{In the formula}: \ c_{j,n+4} = \frac{-c_{j,n}}{(n+4)(n+3)(n+2)(n+1)} \tag{5.9}$$

Among them, $n \geq 0$; $c_j = \begin{cases} 1, \ j = i \\ 0, \ j \neq i \end{cases} \quad (j, i = 0, 1, 2, 3)$.

If Eq. (5.8) is substituted into Eq. (5.6), then:

$$w_1 = a_0 W_0(\overline{x}) + \frac{a_1}{\alpha_1}W_1(\overline{x}) + \frac{a_2}{\alpha_1^2}W_2(\overline{x}) + \frac{a_3}{\alpha_1^3}W_3(\overline{x}) \tag{5.10}$$

In the formula: $\overline{x} = \alpha_1 x$; $W_j(\overline{x}) = \sum_{n=0}^{\infty} c_{j,n}(\overline{x})^n$. Once, twice and three derivatives of w1 can be obtained respectively:

$$\begin{bmatrix} \dfrac{\theta_1}{\alpha_1} \\ -\dfrac{M_1}{EI\alpha_1^3} \\ -\dfrac{Q_1}{EI\alpha_1^3} \end{bmatrix} = \begin{bmatrix} W_0'(\overline{x}) & W_1'(\overline{x}) & W_2'(\overline{x}) & W_3'(\overline{x}) \\ W_0''(\overline{x}) & W_1''(\overline{x}) & W_2''(\overline{x}) & W_3''(\overline{x}) \\ W_0^{*''}(\overline{x}) & W_1^{*''}(\overline{x}) & W_2^{*''}(\overline{x}) & W_3^{*''}(\overline{x}) \end{bmatrix} \begin{bmatrix} a_0 \\ \dfrac{a_1}{\alpha_1} \\ \dfrac{a_2}{\alpha_1^2} \\ \dfrac{a_3}{\alpha_1^3} \end{bmatrix} \tag{5.11}$$

In the formula:

$$W_j'(\overline{x}) = \sum_{n=1}^{\infty} nc_{j,n}(\overline{x})^{n-1}; \ W_j''(\overline{x}) = \sum_{n=2}^{\infty}(n-1)c_{j,n}(\overline{x})^{n-2};$$

$$W_j'''(\overline{x}) = \sum_{n=3}^{\infty} n(n-1)(n-2)c_{j,n}(\overline{x})^{n-3}; \ W_j^{*''}(\overline{x}) = W_j'''(\overline{x}); \quad j = 0, 1, 2, 3$$

The vertical displacement, inclination Angle, bending moment and shear force at the midpoint of the beam layer are recorded as

$$w_0, \theta_0, M_0, Q_0$$

, and, according to Eqs. (5.7) and (5.11), when X = 0, the following equation can be obtained:

$$\begin{cases} a_0 = w_0 \\ a_1 = \theta_0 \\ a_2 = -\dfrac{M_0}{2EI} \\ a_3 = -\dfrac{Q_0}{6EI} \end{cases} \tag{5.12}$$

As a result, the structure of the arbitrary point displacement and internal force can be w_0, θ_0, M_0, Q_0:

$$\begin{bmatrix} w_1 \\ \dfrac{\theta_1}{\alpha_1} \\ -\dfrac{M_1}{EI\alpha_1^2} \\ -\dfrac{Q_1}{EI\alpha_1^3} \end{bmatrix} = A_1 \begin{bmatrix} w_0 \\ \dfrac{\theta_0}{\alpha_0} \\ -\dfrac{M_0}{EI\alpha_1^2} \\ -\dfrac{Q_0}{EI\alpha_1^3} \end{bmatrix} \tag{5.13}$$

Among them: $A_1 = \begin{bmatrix} W_0(\overline{x}) & W_1(\overline{x}) & \frac{W_2(\overline{x})}{2} & \frac{W_3(\overline{x})}{6} \\ W_0'(\overline{x}) & W_1'(\overline{x}) & \frac{W_2'(\overline{x})}{2} & \frac{W_2'(\overline{x})}{6} \\ W_0''(\overline{x}) & W_1''(\overline{x}) & \frac{W_2''(\overline{x})}{2} & \frac{W_3''(\overline{x})}{6} \\ W_0^{*''}(\overline{x}) & W_1^{*''}(\overline{x}) & \frac{W_2^{*''}(\overline{x})}{2} & \frac{W_3^{*''}(\overline{x})}{6} \end{bmatrix}$.

As shown in Fig. 5.3, while analyzing the corresponding second paragraph, namely $S/2 \le x \le (S/2 + d)$, the corresponding local coordinate system $t_2 - z[t_2 = x - l_1, (l_1 = S/2)]$ can be set up. Similarly, take any micro element within the range for calculation and analysis. And the formula for calculating the internal force and displacement of the building can be obtained:

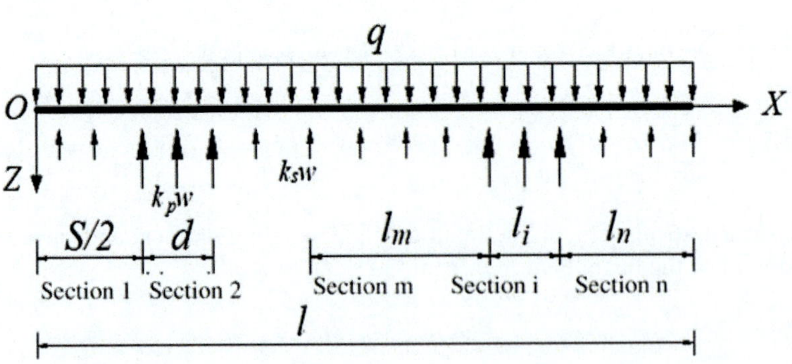

Fig. 5.3 Division of building analysis units

$$\begin{bmatrix} w_2 \\ \dfrac{\theta_2}{\alpha_2} \\ -\dfrac{M_2}{EI\alpha_2^2} \\ -\dfrac{Q_2}{EI\alpha_2^3} \end{bmatrix} = A_2 \begin{bmatrix} w_{2,0} \\ \dfrac{\theta_{2,0}}{\alpha_2} \\ -\dfrac{M_{2,0}}{EI\alpha_2^2} \\ -\dfrac{Q_{2,0}}{EI\alpha_2^3} \end{bmatrix} \tag{5.14}$$

where, the calculation formula of A_2 is the same as that of A_1 respectively, and we just substitute the $t_2 = x - l_1$ for the x, in the calculation, $\alpha_2 = \sqrt[4]{\frac{Bk_p}{EI}}$ for the α_1, and $L_2 = \sum_{j=2}^{n} l_j$ for L_1 in the calculation of$c_{j,n}$. The subscript (2, 0) corresponding to the point$t_2 = 0$.

According to the continuity conditions of the equation $x = l_1 = S/2$, that is $\begin{cases} w_{2,0} = w_{1,l_1} \\ \theta_{2,0} = \theta_{1,l_1} \\ M_{2,0} = M_{1,l_1} \\ Q_{2,0} = Q_{1,l_1} \end{cases}$ The subscript $(1, l_1)$ corresponds to the point $x = l_1$, and $w_{1,l_1}, \theta_{1,l_1}, M_{1,l_1}, Q_{1,l_1}$ can be determined by Eq. (5.13), therefore,

$$\begin{bmatrix} w_2 \\ \dfrac{\theta_2}{\alpha_2} \\ -\dfrac{M_2}{EI\alpha_2^2} \\ -\dfrac{Q_2}{EI\alpha_2^3} \end{bmatrix} = \overline{A}_2 \left(A_1^* \begin{bmatrix} w_{2,0} \\ \dfrac{\theta_{2,0}}{\alpha_2} \\ -\dfrac{M_{2,0}}{EI\alpha_2^2} \\ -\dfrac{Q_{2,0}}{EI\alpha_2^3} \end{bmatrix} \right) \tag{5.15}$$

In the formula: $\overline{A}_2 = A_2 \begin{bmatrix} 1 & 0 & 0 & 0 \\ 0 & \frac{\alpha_1}{\alpha_2} & 0 & 0 \\ 0 & 0 & \frac{\alpha_1^2}{\alpha_2^2} & 0 \\ 0 & 0 & 0 & \frac{\alpha_1^3}{\alpha_2^3} \end{bmatrix}$; A_1^* is the value of A_1 when $x = l_1$.

If the left-hand side of section m, that is during $x = x_{m-1} = \sum_{i=1}^{m-1} l_i$, which section acts under uniform load q_j, similarly, the m element segment satisfies:

$$\begin{bmatrix} w_m \\ \frac{\theta_m}{\alpha_m} \\ -\frac{M_m}{EI\alpha_m^2} \\ -\frac{Q_m}{EI\alpha_m^3} \end{bmatrix} = A_m \begin{bmatrix} w_{m,0} \\ \frac{\theta_{m,0}}{\alpha_2} \\ -\frac{M_{m,0}}{EI\alpha_m^2} \\ -\frac{Q_{m,0}}{EI\alpha_m^3} \end{bmatrix} \tag{5.16}$$

According to the continuity conditions of the equation $x = \sum_{i=1}^{m-1} l_i$, that is,

$\begin{cases} w_{m,0} = w_{m-1,l_{m-1}} \\ \theta_{m,0} = \theta_{m-1,l_{m-1}} \\ M_{m,0} = M_{m-1,l_{m-1}} \\ Q_{m,0} = Q_{m-1,l_{m-1}} - ql_j \end{cases}$, therefore:

$$\begin{aligned} \begin{bmatrix} w_m \\ \frac{\theta_m}{\alpha_m} \\ -\frac{M_m}{EI\alpha_m^2} \\ -\frac{Q_m}{EI\alpha_m^3} \end{bmatrix} &= A_m \begin{bmatrix} 0 \\ 0 \\ 0 \\ \frac{P_j}{EI\alpha_m^3} \end{bmatrix} + \overline{A}_m \begin{bmatrix} w_{m-1,l_{m-1}} \\ \frac{\theta_{m-1,l_{m-1}}}{\alpha_{m-1}} \\ -\frac{M_{m-1,l_{m-1}}}{EI\alpha_{m-1}^2} \\ -\frac{Q_{m-1,l_{m-1}}}{EI\alpha_{m-1}^3} \end{bmatrix} \\ &= A_m \begin{bmatrix} 0 \\ 0 \\ 0 \\ \frac{P_j}{EI\alpha_m^3} \end{bmatrix} + \overline{A}_m \left(\overline{A}_m^* \begin{bmatrix} w_{m-1,0} \\ \frac{\theta_{m-1,0}}{\alpha_{m-1}} \\ -\frac{M_{m-1,0}}{EI\alpha_{m-1}^2} \\ -\frac{Q_{m-1,0}}{EI\alpha_{m-1}^3} \end{bmatrix} \right) \end{aligned} \tag{5.17}$$

where, $\overline{A}_m = A_m \begin{bmatrix} 1 & 0 & 0 & 0 \\ 0 & \frac{\alpha_{m-1}}{\alpha_m} & 0 & 0 \\ 0 & 0 & \frac{\alpha_{m-1}^2}{\alpha_m^2} & 0 \\ 0 & 0 & 0 & \frac{\alpha_{m-1}^3}{\alpha_m^3} \end{bmatrix}$, the calculation formula of A_m is the same as that of A_1 respectively, and we just substitute the $t_m = x - \sum_{j=1}^{m-1} l_j$ for the x in the calculation, $\alpha_m = \sqrt[4]{\frac{Bk_p}{EI}}$ (In the case of inter-pile soil unit, $\alpha_m = \sqrt[4]{\frac{Bk_s}{EI}}$) for the α_1, and

$L_m = \sum_{j=m}^{n} l_j$ for L_1 in the calculation of $c_{j,n}$. Substitute $t_{m-1} = l_{m-1}$ into A_{m-1} and you get A^*_{m-1}. By analogy, the expressions of the deformation, inclination, bending moment and shear force of the building within the range of $0 \le x \le l$ can be obtained:

$$\begin{bmatrix} w_i \\ \dfrac{\theta_i}{\alpha_i} \\ -\dfrac{M_i}{EI\alpha_i^2} \\ -\dfrac{Q_i}{EI\alpha_i^3} \end{bmatrix} = \overline{A}_i \overline{A}^*_{i-1} \cdots\cdots \overline{A}^*_2 \overline{A}^*_1 \begin{bmatrix} w_0 \\ \dfrac{\theta_0}{\alpha_1} \\ -\dfrac{M_0}{EI\alpha_1^2} \\ -\dfrac{Q_0}{EI\alpha_1^3} \end{bmatrix} \tag{5.18}$$

Thus, the internal force and displacement of the building in any range can be expressed by w_0, θ_0, M_0 and Q_0. w_0, θ_0, M_0 and Q_0 are based on the following boundary conditions: if the two ends of the building are free, then

$$\begin{cases} M|_{x=0} = 0 \\ Q|_{x=0} = 0 \end{cases} \quad \begin{cases} M|_{x=l} = 0 \\ Q|_{x=l} = 0 \end{cases}$$

Then find a particular solution w^* of the nonhomogeneous differential Eq. (5.4), then the solution of the differential Eq. (5.4) is the general solution plus the particular solution. Since the Eq. (5.4) cannot derive the theoretical analytical solution, 1stOpt software can be used for numerical solution.

5.2.2 *Theoretical Analysis*

Basic conditions for calculation and analysis: The model as shown in Fig. 5.4 was established. The bending stiffness *EJ* of the foundation beam was 1280 MN·m^2, the length of the building wall l was 20 m, the vertical load q of the building on the foundation was 200 kN/m^2, and the soft soil foundation bed coefficient k_s was 15000 kN/m^3, short pile stiffness coefficient K_p take 200 × 10^3 kN/m, the average stiffness coefficient of pile their section $k_p = k_s + \frac{K_p}{dB} = 415000$ kN/m^3, length of short pile is 5 m, and diameter is 0.5 m. The diameter of the shield tunneling machine is 6.34 m, the length is 6 m, the inner and outer diameters of the lining are 5.50 m and 6.20 m respectively, the width is 1.0 m, the depth of the tunnel axis h is 9.1 m, the grouting rate is 200% and the grouting loss is 65%.

By substituting the parameters into the synergistic differential equation and using 1stOpt nonlinear software for numerical calculation, the building settlement, tilt

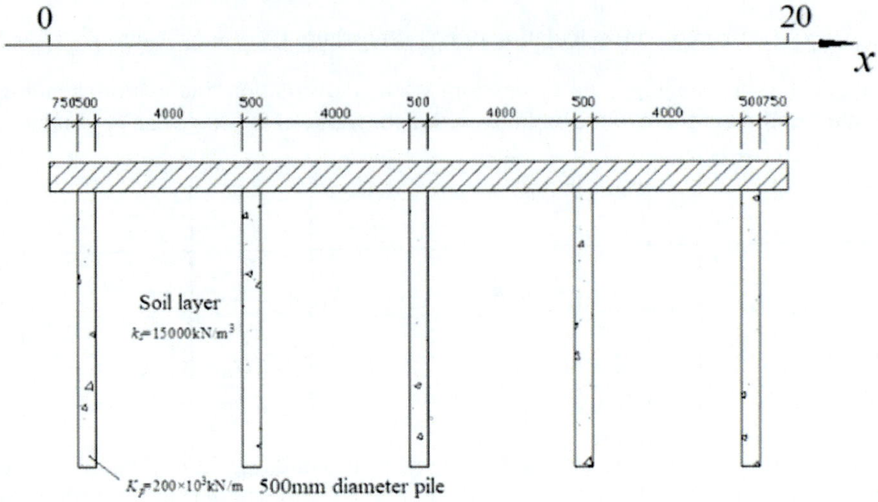

Fig. 5.4 Building model

curve, bending moment and shear distribution curve and additional stress change curve of the building are obtained.

This book selects 6 situations with obvious rules for comparison (50, 10, 0, −10, −20, −30 and −50 m from the excavation face to the left end of the building), as shown in Fig. 5.5.

It can be seen from Fig. 5.5a that when the shield does not reach the building and is still far away, due to the presence of pile foundation, the initial settlement in the middle of the building is about 3.5 mm, while the settlement at both ends is slightly smaller. When the shield approaches the building, the left end of the building (closest to the excavation face) begins to sink. As the shield approaches, arrives, passes through and leaves the building, the settlement of the building develops from the left end to the right end. When the shield excavation face is located under the left half of the building, the settlement speed of the left end of the building is very fast. When the shield excavation face is 10 m away from the building, the settlement of the right end of the building develops very fast. When the shield is far away from the building, the settlement difference between the left and right ends of the building is almost zero, and the settlement in the middle reaches 17.8 mm. It can be seen from Fig. 5.5b that when the building is not affected by the shield, there is a slight slope at both ends. From the curves j = −10 m and j = −20 m, it can be seen that when the shield is located under the right half of the building, the overall slope of the building is larger, and it is the largest when the excavation face is just separated from the building. The development of the settlement in the middle of the building has been relatively uniform in the process of shield tunneling.

It can be seen from Fig. 5.5c that when the building is not affected by shield tunneling, the bending moment with wave-like distribution already exists on the

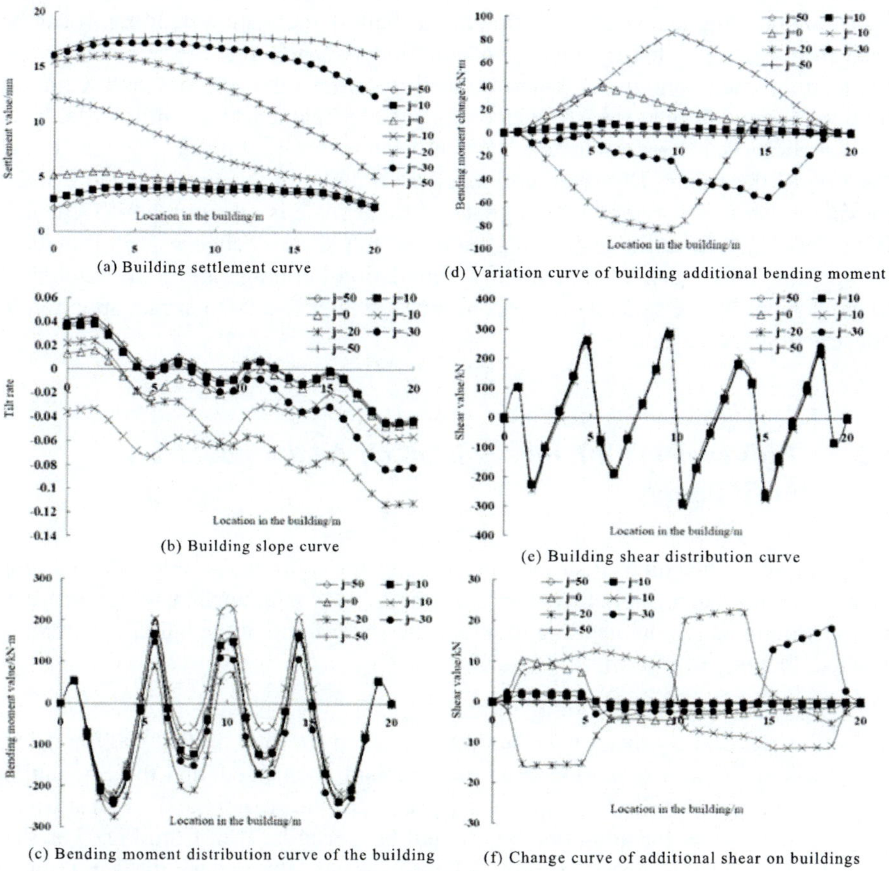

(a) Building settlement curve
(b) Building slope curve
(c) Bending moment distribution curve of the building
(d) Variation curve of building additional bending moment
(e) Building shear distribution curve
(f) Change curve of additional shear on buildings

Fig. 5.5 Building deformation and internal force change curve

foundation beam. The peak of positive bending moment is located at the center of the pile, while the negative bending moment is located at the center of the middle section of two adjacent piles, which is obviously different from the shallow foundation. During the whole process of the shield approaching, crossing and leaving the building, the additional bending moment near the two ends of the building has little change, while the middle part has great change. It can be seen from Fig. 5.5e that when the building is not affected by shield tunneling, shear force has been distributed serrated on the foundation beam. The peak of positive shear force is located in the center of the pile, while the negative shear force is located in the center of the middle section of two adjacent piles. Under the influence of shield tunneling, the additional shear force on the foundation beam has little change.

Figure 5.5d and f show the variation of bending moment and shear force compared with $j = -50$ m, respectively. It can be seen from Fig. 5.5d that, under the influence of shield tunneling, the distribution law of additional bending moment variation on short pile building is consistent with that on shallow foundation building. It can be seen from Fig. 5.5f that the change of additional shear force on the beam on the foundation is not continuous, but is divided into four relatively obvious segments by piles. The additional shear values on the smaller segments have little difference, but the additional shear values on different segments vary with the changes in the location of the shield. However, compared with the initial value (j = −50 m), the maximum variation of bending moment and shear force is no more than 1/3, which is far less than the ratio of bending moment and shear change value of shallow foundation. This indicates that shallow foundation buildings are more vulnerable to damage under the same working conditions, which is easy to attract attention in construction and design.

5.2.3 *Influence of Different Factors on Additional Stress of Buildings*

When the shield machine undergoes the building, the additional stress of the building is in dynamic change, and other working conditions remain unchanged. The case of the maximum additional bending moment and shear force of the building is mainly studied, that is, $j = -10$ m.

(1) Influence of flexural rigidity of buildings
The flexural rigidity of the building is taken as 1500, 1280, 1000 and 750 MN·m^2, other calculation conditions remain unchanged, and the calculation results are shown in Fig. 5.6. It can be seen that the greater the flexural stiffness is, the greater the additional stress will be. It can be seen from Fig. 5.6a that the greater the flexural rigidity of the building, the greater the variation of additional bending moment. This is because the greater the flexural rigidity, the more difficult the building will be to deform with the surface deformation. The greater the accumulated energy in the building, the greater the additional bending moment will be. In addition, with the increase of the flexural rigidity of the building, the increase of additional bending moment becomes slow.It can be seen from Fig. 5.6b that the change rule of shear force is similar to bending moment, but there is a sudden change in the existing position of pile foundation. In the process of shield tunneling, the connection between pile foundation and ground beam is more prone to stress concentration and damage, which should be paid attention to during construction.

(2) Influence of stiffness coefficient of short pile
The stiffness coefficient of short pile is taken as 250, 200, 150 and 100kN/m respectively, other calculation conditions remain unchanged, and the calculation results are shown in Fig. 5.7. It can be seen from the figure that the additional stress of the building is not significantly affected by the stiffness

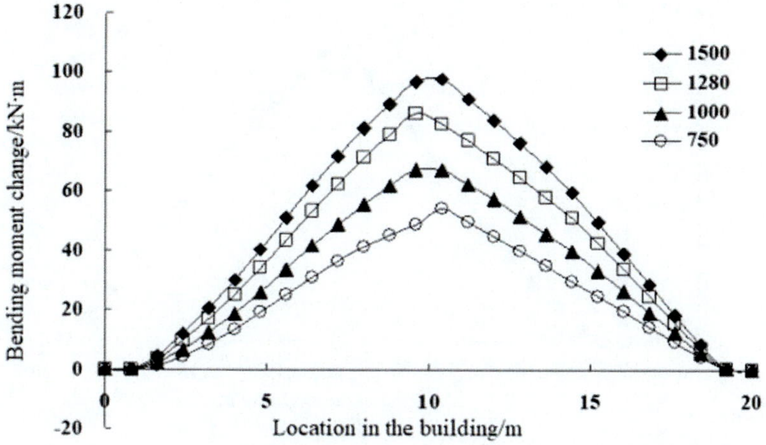

(a) The bending stiffness of the building is not the same as the additional bending moment distribution curve

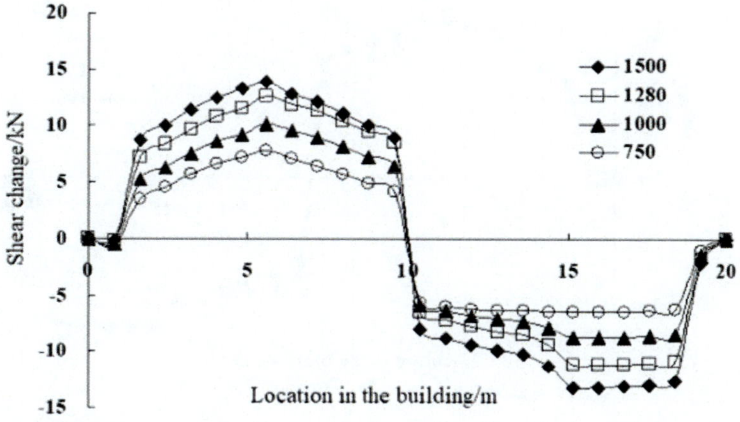

(b) Additional shear distribution curve of buildings with different flexural stiffness

Fig. 5.6 Additional stress distribution curve when flexural stiffness is different

coefficient of the short pile, which means that only increasing the stiffness of the pile itself is not particularly useful for resisting the deformation caused by shield tunneling. It is worth noting in the preliminary design that there is no need to blindly increase the concrete strength of the pile.

(3) Influence of grouting rate

Other calculation conditions remain unchanged, the grouting rate is respectively 125%, 150%, 175% and 200%. It can be seen in the Fig. 5.8 for the calculation results. And it can be seen from the figure that the higher the grouting rate, the less the additional stress of the building will be under the

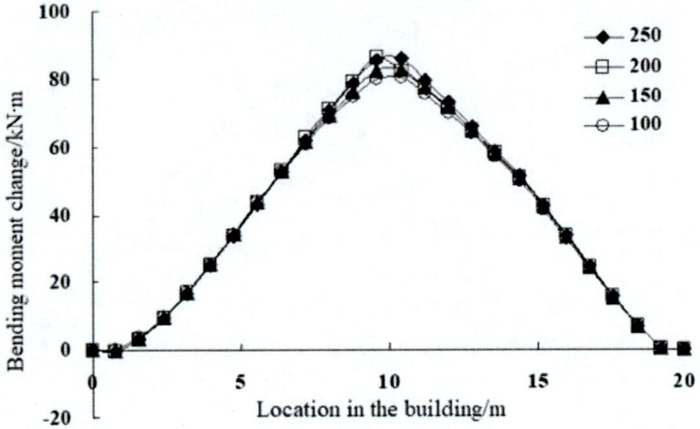

(a) Add bending moment distribution curve when the stiffness coefficient of short pile is different

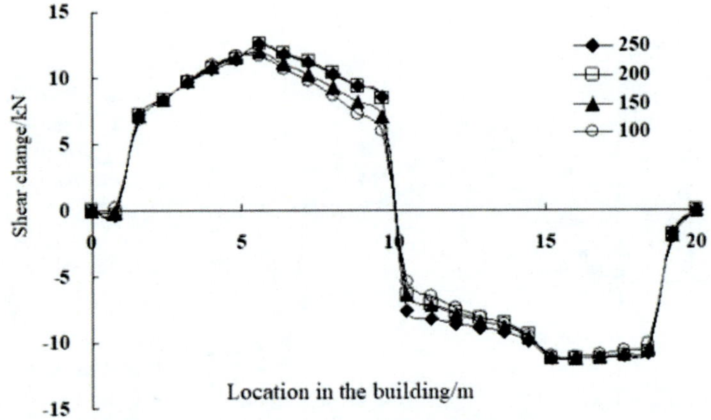

(b) Additional shear distribution curve with different stiffness coefficient of short pile

Fig. 5.7 Additional stress distribution curve when stiffness coefficient of short pile is different

condition that the grouting rate will not cause uplift of the ground. This is because the increase of the grouting rate can better fill the gap of the shield tail and reduce the deformation of the ground and the building, so as to avoid great impact on the building above the tunnel.

(4) Influence of grouting loss

Other calculation conditions remain unchanged, the grouting loss is taken as 75%, 60%, 60% and 55% respectively, and the calculation results are shown in Fig. 5.9. It can be seen from the figure that the smaller the grouting loss is, the smaller the additional stress of the building will be. This is the same

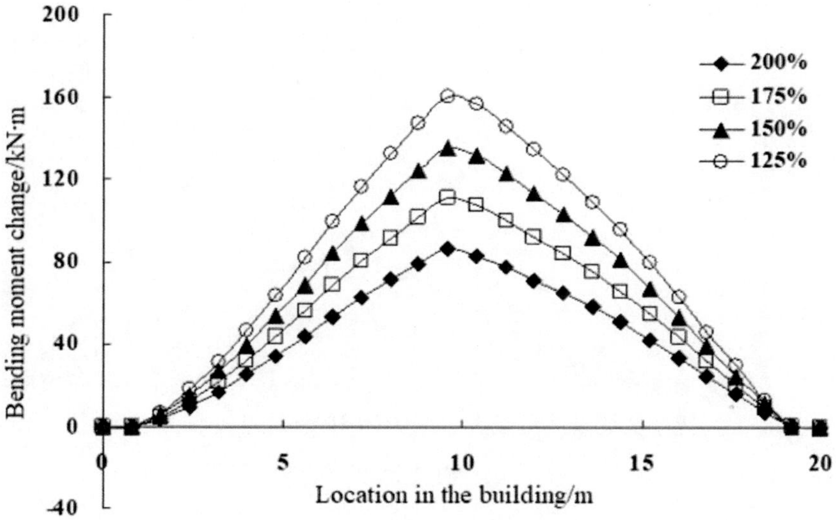

(a) Add bending moment distribution curve when grouting rate is different

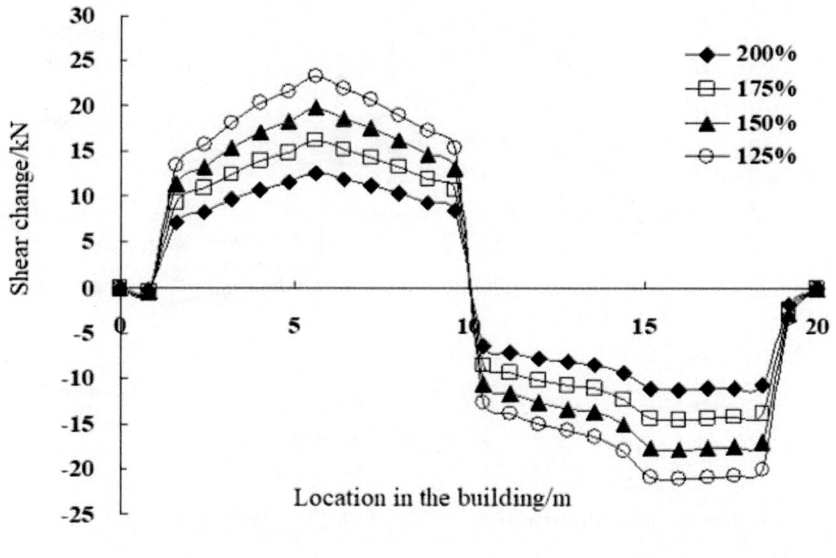

(b) Additional shear distribution curve with different grouting rates

Fig. 5.8 Additional stress distribution curve when grouting rate is different

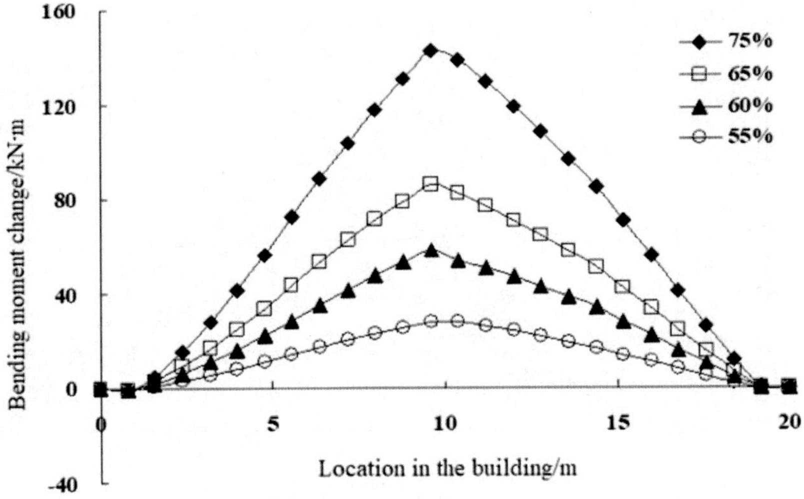

(a) Add bending moment distribution curve when grouting loss rate is different

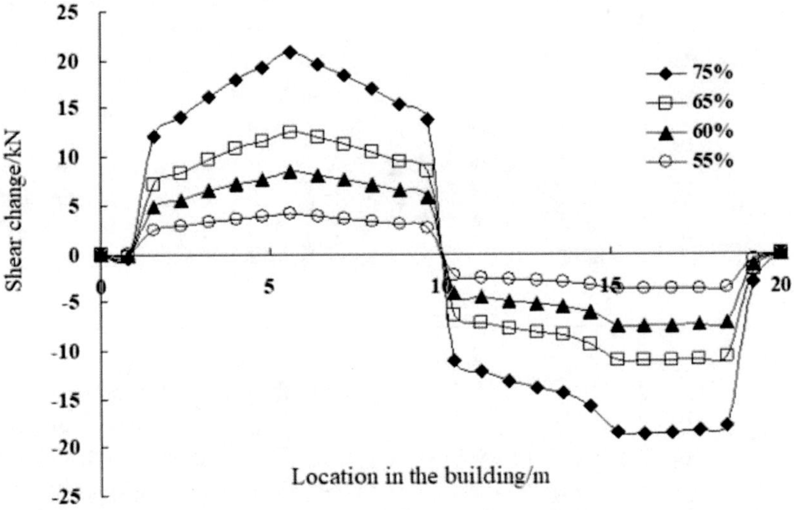

(b) Additional shear distribution curve when the grouting loss rate is different

Fig. 5.9 Additional shear distribution curve when the grouting loss rate is different as the influence mechanism of the grouting rate. Reducing the grouting loss can make the grouting effect behind the wall better, so as to effectively control the settlement of the ground and the building, and reduce the tilt and bending deformation of the building.

5.3 Theoretical Research on the Synergy Model of Shear and Bending Beams Based on Elastic Foundation

5.3.1 Establishment of Calculation Model

The double-coordinate system in the previous section is still adopted, and the building is a frame structure, as shown in Fig. 5.10.

The superstructure of the short pile building is simplified as a shear bent beam on the elastic foundation, and the theoretical differential equation of the synergistic action of the building foundation, pile foundation and structure above the shield tunnel is obtained in the same way as the derivation in Sect. 4.3:

$$
\begin{aligned}
& EJ\frac{\mathrm{d}^4 w(x)}{\mathrm{d}x^4} - (GF + g)\frac{\mathrm{d}^2 w(x)}{\mathrm{d}x^2} + k' w(x) \\
& = q + \frac{k'(V_{loss2} - V_{loss3})}{2\pi h}\left[1 - \frac{j + x + L}{\sqrt{(j + x + L)^2 + h^2}}\right]
\end{aligned}
\tag{5.19}
$$

where, *EJ* is the bending stiffness of building foundation beam, the unit is kN·m^2; *GF* is the vertical shear stiffness of the frame, in unit kN; *g* is the constrained line stiffness of the bottom column end, in unit kN; *q*(*x*) is the vertical load of the building acting on the foundation, in unit kN/m^2.

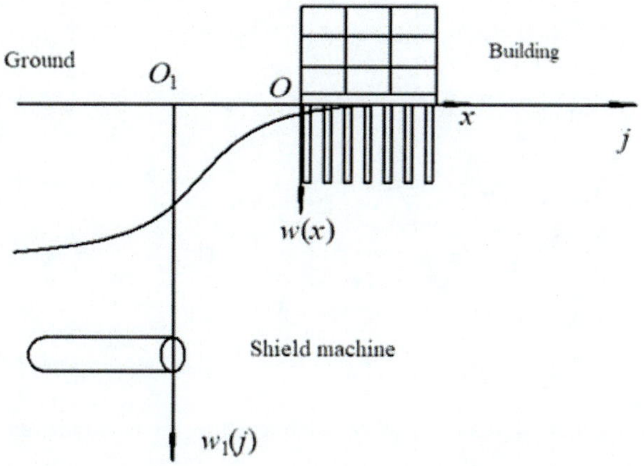

Fig. 5.10 Building and surface settlement coordinate system

5.3.2 Calculation and Analysis of Examples

The example is shown in Sect. 4.3.

1stOpt software is used for numerical calculation to obtain the settlement curve of the building, the additional stress distribution curve of the bottom beam and the variation curve of the internal force of the beam and column, as shown in Fig. 5.11.

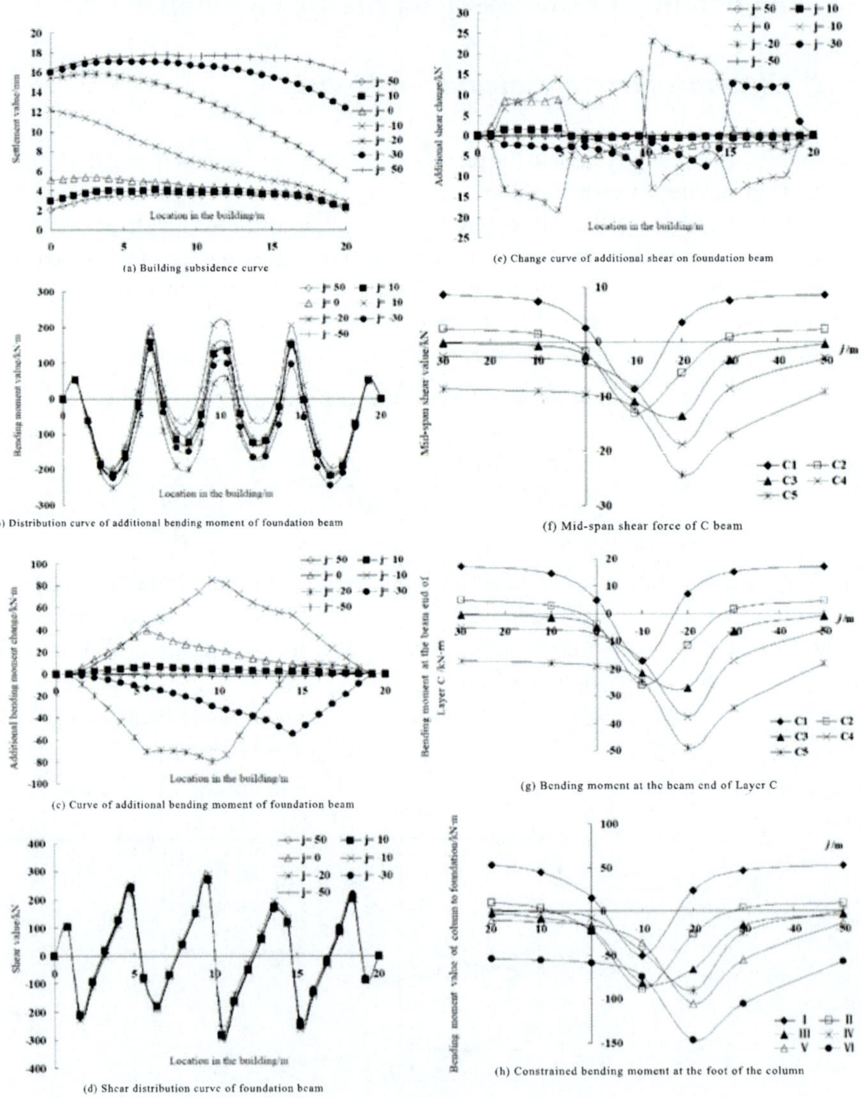

Fig. 5.11 Variation curve of building deformation and internal force

It can be concluded from Fig. 5.11a that when the building is not affected by shield tunneling, the whole building has been cut into the foundation to a certain depth, which is related to the selection of parameters such as pile stiffness, foundation bed coefficient and vertical load, because the building needs foundation reaction to balance the load of the whole building. The settlement of the building is inclined to one side as a whole, and the variation of the settlement value from the left end to the right end is uniform, and the influence of shield tunneling on the settlement of the pile foundation building is less than that of the shallow foundation building.

The additional bending moment and shear force of any section of the building vary greatly during the process of shield tunneling, as shown in Fig. 5.11b and d. Due to the existence of pile foundation, bending moment and shear force have been generated before shield tunneling. When the building is deformed, the additional bending moment and shear force inside the building begin to increase. When the excavation surface reaches the right under the building's approximately 1/2 position (the building is located at the place with the maximum negative and positive curvature of the surface subsidence curve), the bending moment value in the middle of the building reaches the maximum. Finally, when the settlement of the building is stable, the change value of the additional bending moment of the building becomes smaller. No matter where the excavation face is located, the building always has positive and negative bending moment and shear force, which are obviously different from the shallow foundation building. Figure 5.11c and e show the variation of bending moment and shear force compared with $j = -50$ m, respectively. The maximum positive and negative variation of bending moment and shear force are still at the position of $j = -10$ m and $j = -20$ m, which should be paid attention to in construction and design.

As shown in Fig. 5.11f–h, the internal forces of the beams and columns of each layer of the frame gradually increase with the approach of the shield, and the change trend during the process of crossing the building is more gentle than that of the shallow foundation, which indicates that the constraint of the pile foundation plays a greater role in balancing. When the excavation face is located directly under the building or is about to pass through the building completely, the internal force of the beam and column at each position is close to the maximum value. The mid-span shear force of the beam is about −25 kN, the maximum bending moment of the beam end is about −50 kN·m, and the maximum restrained bending moment of the bottom column foot is about −150 kN·m. At the same time, due to the constraint of pile foundation, the maximum position is not exactly the same as that of the shallow foundation frame structure. The maximum internal forces of the right beam and column of the building are smaller than those of the left beam and column, which is closely related to the soil deformation and position caused by shield tunneling.

5.4 Summary

At present, the tunnels of shield tunnel through short pile foundation buildings will still produce deformation and additional stress to a certain extent. In this chapter, the building is simplified as a beam on elastic foundation the theoretical mechanical model of the synergistic action of the foundation, pile foundation and structure above shield tunnel is established, and the deformation and internal force variation law of adjacent short pile foundation building during shield tunneling are analyzed theoretically.

(1) Due to the existence of pile foundation, when the shield is far from the building, the initial settlement of the building is much smaller than that of the shallow foundation building. The settlement of the building develops from the left end to the right end as the shield approaches, arrives, passes through and breaks away from the building. When the shield is located under the right half of the building, the overall inclination of the building is larger and reaches its maximum when the excavation face just separates from the building.
(2) The maximum bending moment value and shear value of short pile building are closely related to the flexural stiffness of building, foundation bed coefficient of foundation, pile stiffness and grouting rate. Improving the construction technology of shield tunnel can effectively reduce the additional stress on the building and reduce the damage to the building. At the same time, strengthening the building itself is also one of the effective means to resist deformation, but increasing the pile stiffness has little influence on reducing the deformation of the building.
(3) Due to the existence of pile foundation, before the shield arrives at the building, there are wave-like bending moments and saw-toothed shear forces distributed on the foundation beam. The peak of positive bending moment and shear force is located in the center of the pile, the while the negative bending moment and shear force are located in the center of the middle section of two adjacent piles of the building, which is obviously different from the shallow foundation. In the whole process of shield approach, crossing and leaving the building, the initial values of additional bending moment and shear force near the two ends of the building have little change, indicating that the effect of pile foundation to resist deformation is much better than that of shallow foundation.
(4) The variation trend of the internal forces of the beams and columns in each layer of the frame is more gentle than that of the shallow foundation, which indicates that the constraint of the pile foundation plays a great role in balancing. At the same time, due to the constraint of pile foundation, the maximum position is not the same as that of the shallow foundation frame structure. The maximum internal forces of the right beam and column of the building are smaller than those of the left beam and column, which is closely related to the soil deformation and position caused by shield tunneling. Therefore, in the process of

construction, it is still necessary to strengthen the monitoring of settlement difference and tilt rate between the head and the tail of short pile buildings, so as to better control the influence of shield tunneling on the internal force of frame structure of adjacent pile foundation.

The synergy model of building foundation, pile foundation and structure proposed in this book is a planar model, and it only considers the working condition of short pile foundation, not the working condition of long pile and supporting pile foundation. The model can be further improved to carry out in-depth analysis from the three-dimensional close construction state of different working conditions.

References

1. Sun Y. Research on the mechanical behaviors of masonry building induced by shield tunneling construction. Hangzhou: Zhejiang University;2011.
2. Zhou W. Research on the influence on surrounding buildings induced by tunneling and the face stability of large shield-driven tunnel. Beijing: Tsinghua University;2012.
3. Sun J. Urban environmental geotechnical. Shanghai: Shanghai Science and Technology Press;2005.
4. Wei G, Wei X, Ding Z, et al. Analysis of influence of pipe jacking construction on adjacent piles. Rock Soil Mech. 2006;27(S1):849–54.
5. Wei X, Hong J, Wei G. Analysis of additional load on adjacent pile foundation induced by double-o-tube shield tunnel construction. Rock Soil Mech. 2013;34(3):783–90.
6. Zhao M, Zhang L, Zhao H. Settlement calculation of two-directional reinforced composite foundation. Rock Soil Mech. 2011;32(9):2741–6.

Chapter 6
Study on the Influence and Control Standard of Double Line Shield Tunneling on Adjacent Buildings

6.1 Introduction

In recent years, with the continuous development of China's subway construction, double-line parallel shield construction has become the mainstream form of urban subway tunnel construction. However, shield tunneling will inevitably cause disturbance and deformation of the surrounding soil and cause great harm to the adjacent buildings, especially when double-line parallel tunnels are constructed at the same time, the influence will be superimposed on each other and the formation deformation will be more complex.

At present, scholars at home and abroad have mainly conducted relevant studies on the deformation and disturbance of soil caused by double-line parallel shield, and the research methods mainly include the Peck formula empirical method [1], stochastic medium theory [2], finite element method [3, 4], model test method [5] and analytical solution method [6, 7]. However, there are few studies on the influence of double-line parallel shield construction on the structure of adjacent buildings, especially buildings with shallow foundation frame [8–10], and even fewer studies on the changes of bending moment and shear force of building structures. In particular, double shield tunneling has gradually become an important tunnel construction method due to its advantages such as high construction efficiency, less soil cutting, and the ability to complete the upper and lower lines of subway tunnel in one excavation [11]. About double circle shield construction influence on surrounding environment research focused on the deformation of soil, soil deformation calculation model are equivalent truss model, double circle overlay model and random medium non-uniform convergence model [12–14], and double circle shield construction to adjacent structures especially adjacent shallow foundation frame buildings impact study is few and far between.

Since the tunneling of double-line shield is a dynamic process, it is particularly important to understand the deformation and bending moment changes of the building during the whole process of the shield tunneling through the building. Therefore, this chapter refers to the synergistic model method [15] of mining area, considering the

Z. Ding et al., *Influence of Shield Tunneling on Adjacent Structures and Control Technology*, https://doi.org/10.1007/978-981-19-1134-7_6

complexity of double-line shield construction, the formation loss caused by shield construction is taken as the main cause of formation settlement. The mechanical model of the synergistic action of foundation, foundation and structure of shallow foundation buildings above double-line parallel tunnel and double-circle tunnel is derived. The numerical analysis software 1stOpt is used to solve the problem, and the law of building deformation and internal force change in shield tunneling area is obtained.

6.2 Study on the Influence of Double-Line Shield Tunneling on Adjacent Shallow Foundation Buildings

Considering the complexity of double shield tunnel construction, this section takes the ground loss caused by shield construction as the main reason for ground settlement, and simulates the upper building as a bending beam model on elastic foundation. The problem of building foundation interaction is studied by establishing two mutually independent and interrelated coordinate systems of ground settlement and ground buildings.

6.2.1 Establishment of Joint Action Mechanical Model

(1) Basic assumptions

Two coordinate systems are established as shown in Fig. 6.1. The surface subsidence coordinate system is $w_1(j) - O_1 - j$, and the origin O_1 is established at the surface above the excavation face. The x-axis j points in the same direction as the shield tunnelling machine, and the y-axis) ($w_1(j)$ is the surface subsidence at Point j; Building sink coordinate system is $w(x) - O - x$, The origin is established at the

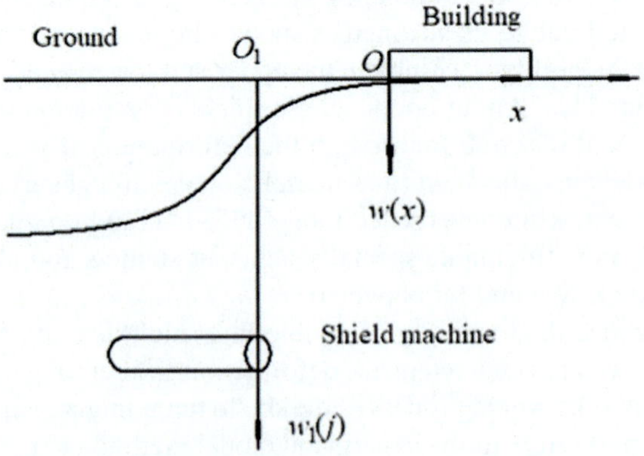

Fig. 6.1 Building and ground subsidence coordinate system

ground of the left end of the building, and the distance from O_1 is j (before the excavation face reaches the building, j is positive; after the excavation face passes through the building and leaves, j is negative), the x-axis is pointing the same way as the j-axis, the subsidence of any point x in the building is $w(x)$, and the corresponding surface point subsidence is $w_1(j + x)$.

1) Foundation model

Treat a building as a beam on an elastic foundation, according to Winkler's elastic foundation theory, the building foundation reaction σ_d is in direct proportion to the foundation settlement value, which can be obtained as follows:

$$\sigma_d(x) = k[w(x) - w_1(j + x)] \tag{6.1}$$

In the formula, $\sigma_d(x)$ is the ground reaction at any point at the bottom of the building, the unit is kN/m^2, and k is the foundation bed coefficient, the unit is kN/m^3.

2) Building model

Differential equation for building bending on elastic foundation:

$$EJ\frac{d^4w(x)}{dx^4} = q - k[w(x) - w_1(j + x)] \tag{6.2}$$

3) Ground deformation model

As shown in Fig. 6.2, soil loss is the main factor causing ground deformation during shield tunnel construction. Vertical displacement formula of the ground presented by Sagaseta [16] is adopted for the surface longitudinal settlement curve:

$$w_1(x) = \frac{a^2}{2}\frac{h}{y^2 + h^2}\left[1 - \frac{x}{\sqrt{x^2 + y^2 + h^2}}\right] \tag{6.3}$$

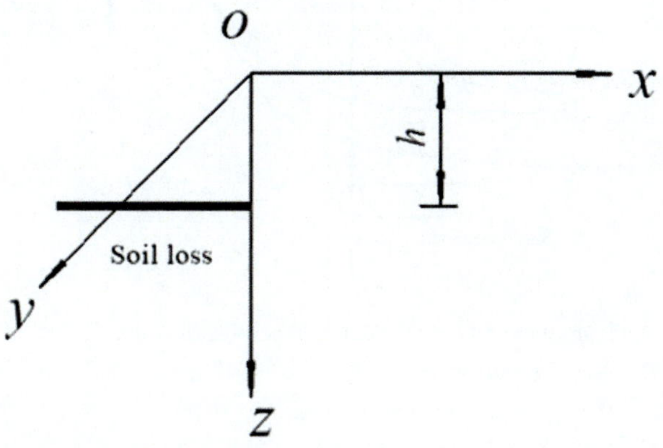

Fig. 6.2 Schematic diagram of soil loss

where: x is the horizontal distance from the digging surface in the driving direction of the shield; y is the transverse horizontal distance from the tunnel axis; a is the loss radius of soil point; h is the depth of tunnel axis, the same as below.

The value of a is related to soil loss, per unit length of the soil loss area is equal to $2\pi a$. The soil loss is calculated by selecting an appropriate percentage of soil loss on the excavation surface, which is usually 0.5–2.5% for clay. Make η percentage for soil loss, then $\pi a^2 = \pi R^2\eta$, that is $a^2 = R^2\eta$, where R is the outer radius of shield [17].

Considering the complexity of double-line shield construction, the land subsidence caused by double-line shield tunneling can be described by applying the superposition principle, that is, the superposition of the longitudinal subsidence curve of the ground [18], as shown in Figs. 6.3 and 6.4, the corresponding land subsidence under buildings caused by double-line shield tunneling can be written

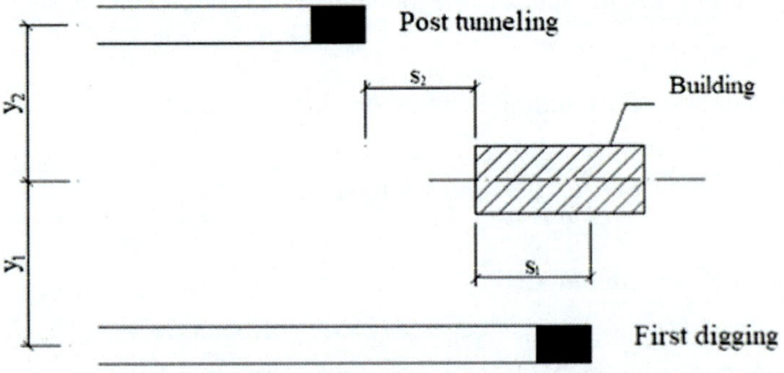

Fig. 6.3 Positional relationship between the double-line tunnel and the building

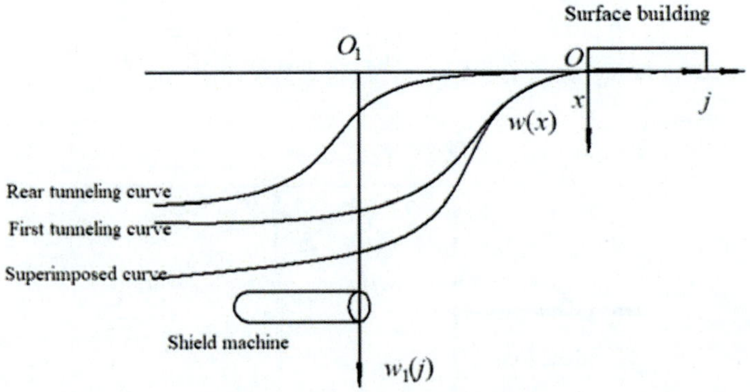

Fig. 6.4 Schematic diagram of surface subsidence superposition curve

$$w_1(i) = \frac{a_1^2}{2}\frac{h_1}{y_1^2 + h_1^2}\left[1 - \frac{i + s_1}{\sqrt{(i + s_1)^2 + y_1^2 + h_1^2}}\right] + \frac{a_2^2}{2}\frac{h_2}{y_2^2 + h_2^2}\left[1 - \frac{i + s_2}{\sqrt{(i + s_2)^2 + y_2^2 + h_2^2}}\right] \tag{6.4}$$

where, s_1 is the longitudinal horizontal distance between the excavation face of the first tunneling tunnel and the left end of the building, and y_1 is the transverse horizontal distance between the axis of the first tunneling tunnel and the building (regardless of the width of the building). Similarly, s_2 is the longitudinal horizontal distance between the excavated face of the rear tunnel and the left end of the building, and y_2 is the transverse horizontal distance between the axis of the rear tunnel and the building.

(2) Synergy model

By substituting Eq. (6.4) into Eq. (6.2), the flexural differential equation of the beam under flexure on the elastic foundation can be obtained:

$$EJ\frac{d^4w(i)}{di^4} + kw(i) = q + \frac{kh_1a_1^2}{2(y_1^2 + h_1^2)}\left[1 - \frac{i + s_1}{\sqrt{(i + s_1)^2 + y_1^2 + h_1^2}}\right] + \frac{kh_2a_2^2}{2(y_2^2 + h_2^2)}\left[1 - \frac{i + s_2}{\sqrt{(i + s_2)^2 + y_2^2 + h_2^2}}\right] \tag{6.5}$$

Equation (6.5) is the differential equation of synergic action of building foundation and strip foundation above the double-line shield tunnel.

On the settlement of research, the beam is simplified to two-dimensional plane bending beam, $y = 0$, longitudinal deformation of the ground above the tunnel axis calculation formula is:

$$EJ\frac{d^4w(i)}{di^4} + kw(i) = q + \frac{ka_1^2}{2h_1}\left[1 - \frac{i + s_1}{\sqrt{(i + s_1)^2 + h_1^2}}\right] + \frac{ka_2^2}{2h_2}\left[1 - \frac{i + s_2}{\sqrt{(i + s_2)^2 + y_2^2 + h_2^2}}\right] \tag{6.6}$$

(3) Others

Refer to Sect. 4.2.1 for boundary conditions, solution process of differential equation and calculation of building bending moment.

6.2.2 Calculation Conditions

It is assumed that the driving direction of the double-line parallel tunnel is consistent with the longitudinal direction of the building, which is located in the middle of the left and right tunnels (that is $y_1 = y_2 = 6$ m), the bending stiffness EJ of foundation beam is 1000MN $\cdot$ m^2, the coefficient of soft soil base bed k is 5000 kN/m^3, the vertical load q of the building on the foundation is 200 kN/m^2, the length of the building wall l is 20 m, soil loss percentage η take 2% (soft soil), the diameter of shield d is 6.2 m, and the depth of tunnel axis h is 9.1 m. The lining has an outer diameter of 6.2 m, an inner diameter of 5.5 m, a ring width of 1.2 m and a thickness of 0.35 m, and C_{50} concrete was used for casting.

In the calculation of this model, the distance between the excavation face of the S_1 shield tunneling machine and the left end of the building. Also, after S_2 of the excavation face of shield machine and the buildings left edge distance, ΔS for shield excavation the distance apart two. We make $\varphi = \Delta S/D$, where the $\Delta S/D$ is for shield construction machine spacing and the ratio of the shield diameter. In the calculation, S_1 takes 30, 10, 0, −10, −20, −30, −50 and −80 respectively to simulate the process of the first shield machine approaching, arriving, passing through and leaving the building; The values of $\Delta S/D$ are 0, 1, 2, 3, 4, 5 respectively to indicate that the shield spacing is simultaneous tunneling, 1*D* successively tunneling, 2*D* successively tunneling, 3*D* successively tunneling, 4*D* successively tunneling, and 5*D* successively tunneling.

By taking the points inside the building: 0 and 20 at the two ends of the plane coordinate axis *X*, $S_1 = 5$ and 15 at the 1/4 and 3/4 positions of the building, and $S_1 = 10$ at the 1/2 position of the building, as the characteristics research points of the building model, the settlement, bending moment and shear force of the building are respectively taken as the research objects.

6.2.3 Case Calculation and Analysis

In the actual construction process, the left and right lines of the double-line shield tunnel are not tunneling at the same time, and in order to reduce the influence of the later shield tunneling on the first shield, the specification requires that the spacing between the two shield tunneling before and after should be more than 100 m apart.

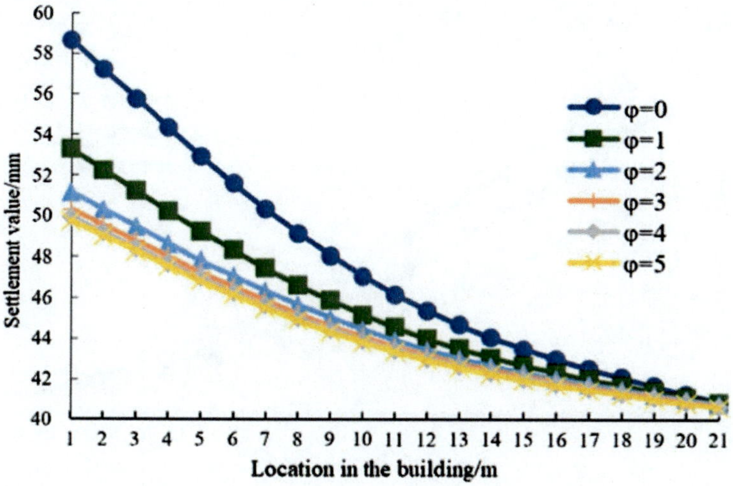

Fig. 6.5 Building settlement distribution curve (when $S_1 = 0$ m)

However, there are still some projects where the distance between the two is relatively small, such as the project case where the shield tunneling after a certain section of Shanghai Rail Transit Line 7 exceeds the shield tunneling before. Therefore, this section mainly explores the situation where the left and right excavation faces are relatively close apart, and analyzes the settlement and internal force variation rules of the building above under different working conditions.

(1) Settlement analysis

1) When the shield tunneling machine arrives at the building (i.e. $S_1 = 0\,\text{m}$)

As shown in Fig. 6.5, when the two shield tunneling machines are driven simultaneously, $\varphi = 0$, $S_1 = S_2 = 0$ m, it can be seen that settlement occurs at the left end of the building (starting from the internal position of 0 m), in comparison with the case of $\varphi = 1 - 5$, the settlement at the left end of the building becomes smaller and smaller until it gradually tends to be 49.6 mm. The settlement from left to right of the building is similar to the conic curve, and the settlement at the right end is small, so when the settlement at the left end is large, the slope of the curve will be larger and the building will be more inclined. In addition, the two shield tunneling and the distance between $1D$ distance drivage, compared to the impact of the distance between two shield under the condition of $1D$ settlement curve is more moderate, and the effect will increase as the distance between two shield (φ value growth), and the performance is more and more obvious. This indicates that under the same construction conditions, when the construction spacing between two shield tunneling machines is larger, the settlement of the building is more gentle when the shield tunneling machine is just passing through, and the damage to the building can be greatly reduced.

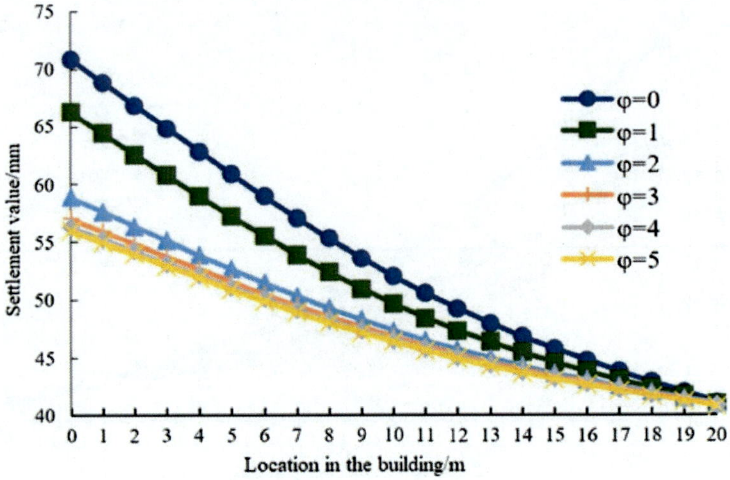

Fig. 6.6 Building settlement distribution curve (when $S_1 = -5$ m)

2) When the shield tunneling machine is located at 1/4 of the building (i.e. $S_1 = -5$ m)

As shown in Fig. 6.6, when the shield tunneling machine has passed through the building, the building settlement continues to occur. The settlement of the left end of the building is intensified, and the slope of the building is the largest when the settlement is the most severe in the case of simultaneous excavation. The settlement curves that are 1*D* apart also have aggravated settlement, and the slope of the curves also varies greatly compared with those that are 2*D* apart. This indicates that the shield tunneling machine also has an impact on the settlement of the building.

3) When the shield tunnelling machine is in the 1/2 position of the building (i.e. $S_1 = -10$ m)

As shown in Fig. 6.7, the settlement of the building is still occurring, and the right end of the tunnelled settlement curve starts to occur at the same time. At this point, the entire settlement curve is similar to a primary function curve, and the tilt of the building tends to be consistent, resulting in the overall left-leaning situation. However, the settlement curve of tunneling more than 3*D* away from each other has a relatively moderate settlement change. The building also leans to the left on the whole, but the inclination is the same, which is much smaller than other conditions.

4) When the shield tunneling machine is in the 3/4 position of the building (i.e. $S_1 = -15$ m)

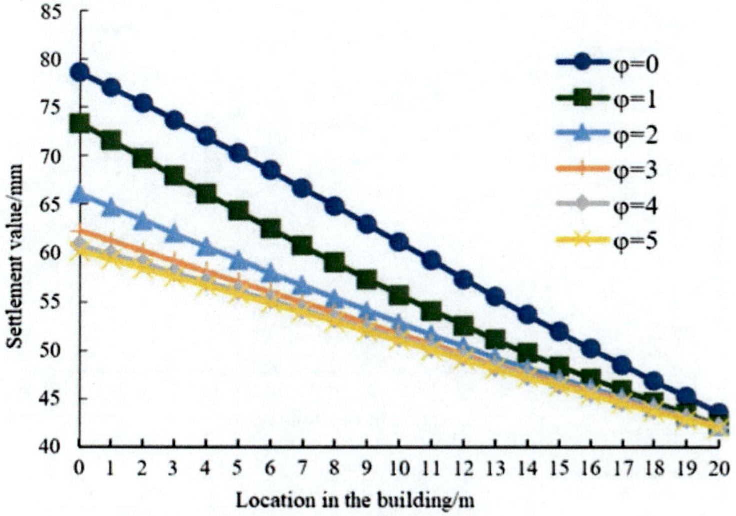

Fig. 6.7 Building settlement distribution curve (when $S_1 = -10$ m)

As shown in Fig. 6.8, the settlement of the building still occurs. In the case of synchronous tunneling, the slope of the settlement curve at the left end of the building has begun to decrease and gradually tends to a stable value. It can be seen that when the excavation surface is about 15 m away from the monitoring point, the settlement of this detection point starts to tend to a fixed value, and the curve gradually stabilises from left to right. The slope of the settlement curve separated by $1D$ and $2D$ tunneling

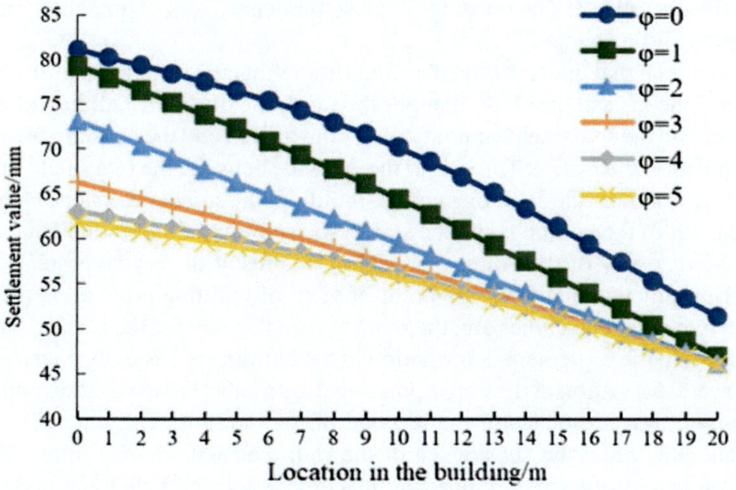

Fig. 6.8 Building settlement distribution curve (when $S_1 = -15$ m)

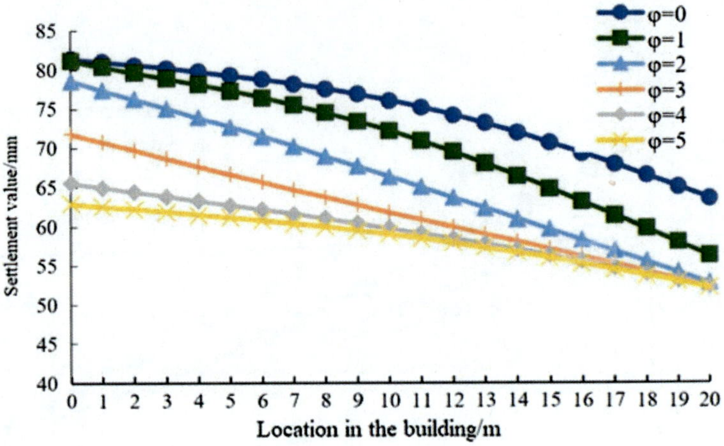

Fig. 6.9 Building settlement distribution curve (when $S_1 = -20$ m)

changes little, and the building turns to the overall left. The settlement curves for tunneling more than 3*D* away from each other changed little.

5) When the shield tunneling machine leaves the building (i.e. $S_1 = -20$ m)

As shown in Fig. 6.9, under the condition of synchronous tunneling, the left end of the settlement curve of the building tends to a stable value, the left end slope decreases to the lowest value, the inclination of the building tends to level off, and the curve tends to be stable from left to right. The settlement curves that are 1*D* apart from each other start to converge to those that are synchronously driven; The slope of the settlement curve apart from 2*D* and 3*D* tunneling changes little, and the building turns to the overall left; The opening of the settlement curves of tunneling more than 4*D* apart has little change.

It can be seen that under the synchronous tunneling, the settlement of the building takes place the fastest, the slope changes the most, the tilt of the building takes place the fastest, and the settlement value stabilizes the fastest, and the building recovers the upright position the fastest. The longer the distance between the two shield tunneling machines, the slower the settlement of the building, the smaller the slope change, the slower the tilt of the building, and the slower the stability of the settlement value, the slower the recovery of the building upright. This result is very consistent with the engineering practice, but from the point of view of building protection, the farther apart the two shield machines are, the more secure they are for the surface buildings.

In addition, the settlement deformation in the building is basically consistent with the surface deformation at the corresponding position: as the excavation approaches the building, the subsidence of the left end of the building will gradually increase and accelerate, while the subsidence of the right end will change little. When the excavation face of the shield tunnelling machine is located below the building, the subsidence volume and velocity of the whole building are similar, which is equivalent to the translation of the building to the ground. It can be seen from Fig. 6.7 that the tilt of the building is the largest when the excavation face is directly below the excavation face. At this time, the difference between the left and right ends of the building is the largest, reaching 35.12 mm. When the excavation face of the shield tunnelling machine leaves the building, the settlement at the left end of the building reaches the maximum and the settlement at the right end continues to develop.

(2) Bending moment analysis

1) When the shield tunneling machine arrives at the building ($S_1 = 0$ m)

As shown in Fig. 6.10, when the excavation face reaches the lower position of the left and right ends of the building, the larger bending moment in the middle of the strip foundation is 99.70 kN · m, and the larger the distance of the shield, the smaller the bending moment is.

2) When the shield tunneling machine is in the 1/4 position of the building ($S_1 = -5$ m)

As shown in Fig. 6.11, the bending moment in the middle of the strip foundation reaches its maximum value, 112.99kN · m, which is the same as that in Sect. 1). The larger the distance between the two shields, the smaller the bending moment in all parts of the building.

3) When the shield tunneling machine is in the 1/2 position of the building ($S_1 = -10$ m)

As shown in Fig. 6.12, when the excavation face is located at the lower part of 1/2 of the building, positive and negative bending moments with similar symmetry appear on the left and right sides of the foundation, and the bending moment in the middle is zero. The moment curves with 1*D* and 2*D* distance apart do not have the same characteristics as other moment curves, but are similar to superposition phenomenon.

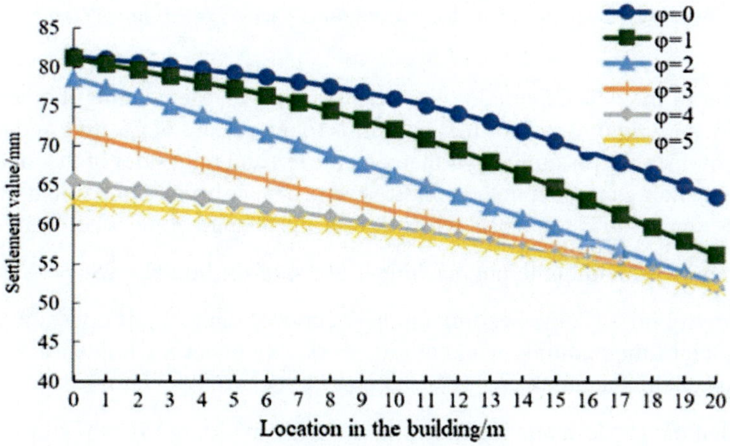

Fig. 6.10 Bending moment distribution curve of the building (when $S_1 = 0$m)

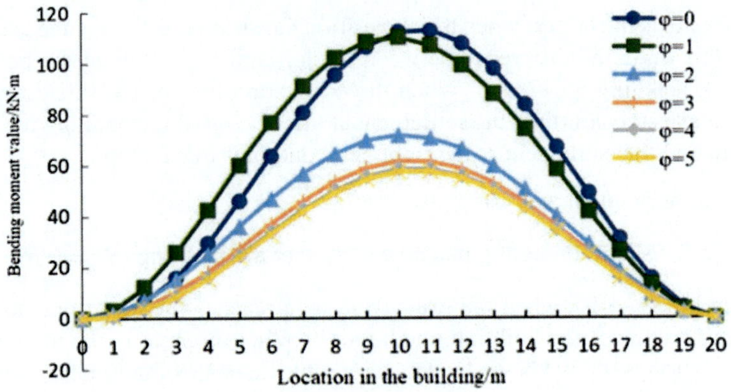

Fig. 6.11 Bending moment distribution curve of the building (when $S_1 = -5\text{m}$)

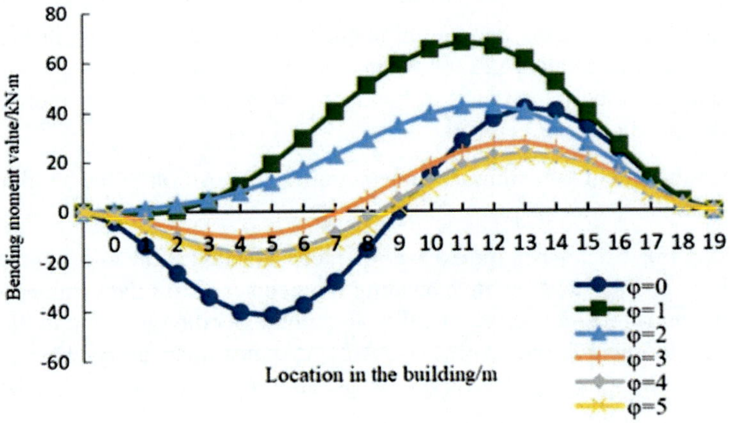

Fig. 6.12 Bending moment distribution curve of beam (when $S_1 = -10\text{m}$)

According to the data calculation, in these two curves, the bending moment curve caused by the shield excavator first conforms to the feature of positive and negative bending moments with similar symmetry on the left and right sides of the foundation, while the bending moment in the middle is zero. This is because the shield excavator after the excavation is too close, which causes the change of the whole curve.

4) When the shield tunneling machine is at 3/4 of the building ($S_1 = -15$ m)

As shown in Fig. 6.13, the bending moment curve of tunneling is opposite to that at 1/4 position of the building, while other curves vary under the influence of the rear tunneling shield machine.

5) When the shield tunneling machine leaves the building ($S_1 = -20$ m)

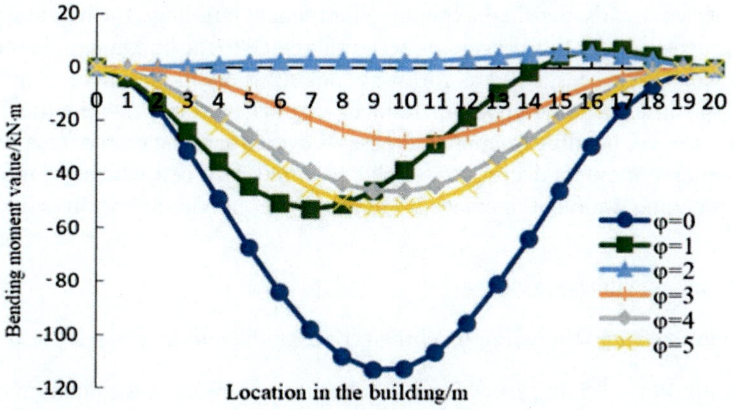

Fig. 6.13 Bending moment distribution curve of beam ($S_1 = -15$ m)

As shown in Fig. 6.14, meanwhile, the bending moment curve of tunneling is opposite to that of the shield tunneling machine when it reaches the building. The middle part of the strip foundation is subject to a large reverse bending moment, which is −99.70 kN · m.

It can be seen that when the shield arrives and leaves, the bending moment in the span is close to the maximum. The passage and departure of shield tunneling machine are two opposite processes. In the whole process of shield tunneling, the foundation beam is subjected to the process of "positive bending moment grows to the maximum—positive bending moment decreases—positive symmetric bending moment appears—reverse bending moment increases—reverse bending moment reaches the maximum". Vertical comparison can obviously show the shield spacing value and the bending moment value develop in opposite directions, the greater the

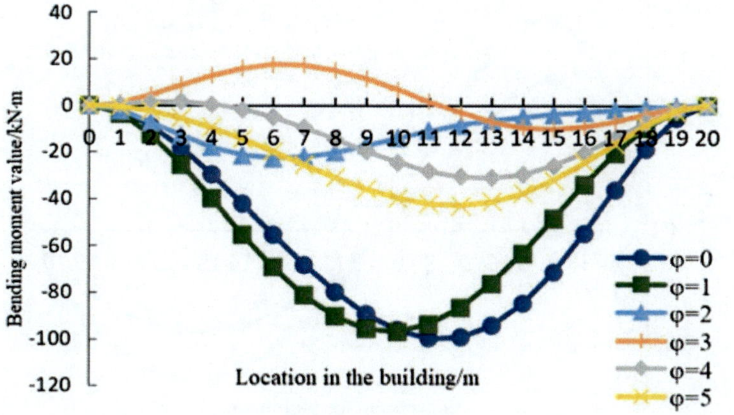

Fig. 6.14 Bending moment distribution curve of beam ($S_1 = -20$ m)

spacing of two shields, the smaller bending moment of buildings, the bending process for buildings is similar to a process of being subjected to the bending moments many times, which can be described as follows: "positive bending moment to the largest growth—positive bending moment reduce—a plus or minus symmetric bending moment—reverse bending moment increase—reverse bending moment maximum—reverse bending moment decreases—a plus or minus symmetric bending moment—positive bending moment increased—reached the maximum positive moment".

(3) Shear force analysis

1) When the shield tunneling machine arrives at the building (i.e. $S_1 = 0$ m)

As shown in Fig. 6.15, the shear force reaches its maximum at the position of about 1/5 and 4/5 of the strip foundation, which is about 17.93 kN at the time of synchronous driving. The larger the shield distance, the smaller the shear force.

2) When the shield tunneling machine is in the 1/4 position of the building (i.e. $S_1 = -5$ m)

As shown in Fig. 6.16, the curve is similar to that when $S_1 = 0$ m, but the curve as a whole tends to move to the right.

3) When the shield tunneling machine is in the 1/2 position of the building (i.e. $S_1 = -10$ m)

As analyzed above, the moment curves with 1*D* and 2*D* distance apart do not have the same characteristics as other shear curves, and similar superposition phenomenon occurs. By checking the data, similarly, the single curve mined in the shear curve is similar to other curves, which is caused by the fact that the shield machine mined after is too close to each other, resulting in the change of the whole curve.

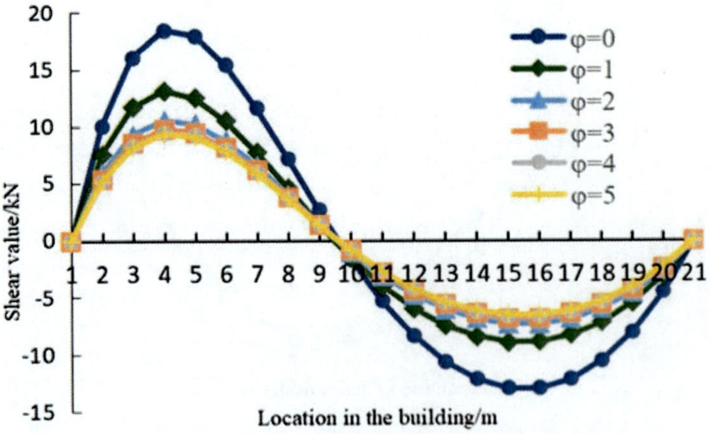

Fig. 6.15 Shear distribution curve of beam (when $S_1 = 0$ m)

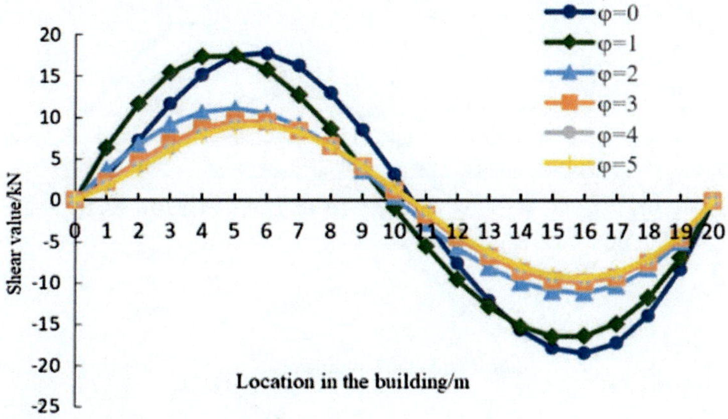

Fig. 6.16 Shear distribution curve of beam (when $S_1 = -5$ m)

As shown in Fig. 6.17, when the excavation face is located directly below 1/2 of the building, positive and negative bending moments appear on the left and right sides of the foundation, while the bending moment in the middle is zero, but the shear force in the middle is the largest. It can be seen that when the shield tunneling through the shallow foundation frame structure, measures to strengthen the central stiffness of the building can achieve the effect of protecting the building.

4) When the shield tunneling machine is in the 3/4 position of the building (i.e. $S_1 = -15$ m)

As shown in Fig. 6.18, the shear curve here is opposite to the shear curve at the 1/4 position of the building, and the varied curves are also affected by the backward driving shield machine.

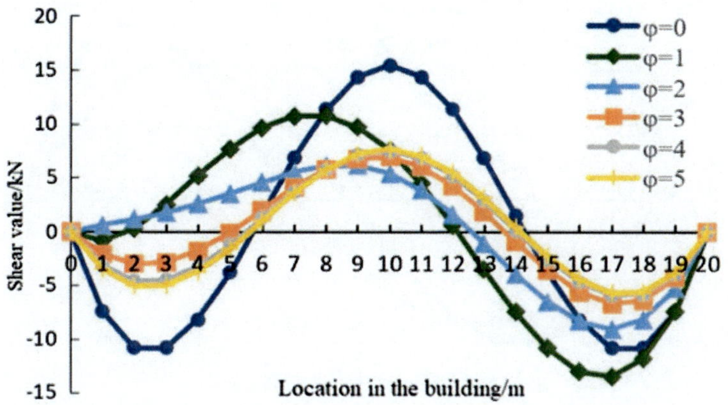

Fig. 6.17 Shear distribution curve of beam (when $S_1 = -10$ m)

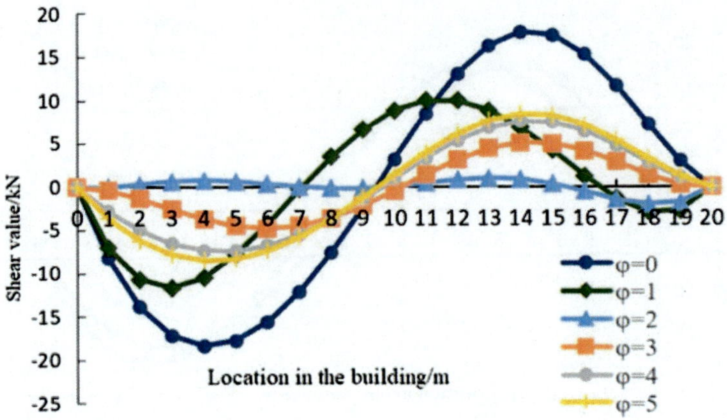

Fig. 6.18 Shear distribution curve of beam (when $S_1 = -15\text{m}$)

5) When the shield tunneling machine leaves the building (i.e. $S_1 = -20$ m)

As shown in Fig. 6.19, at the same time, the shear curve of driving is opposite to that of the shield machine when it arrives at the building, which is −17.92 kN. Moreover, the greater the distance between the shield machines, the more gently the bending moment value changes.

It can be seen that the smaller the distance between the two crossings of the shield tunnelling machine, the shorter the time interval between the two maximum shear forces and bending moments in the building, which is more harmful to the building. Therefore, it is conducive to the safety and stability of the building to set a large space for the double line shield crossing.

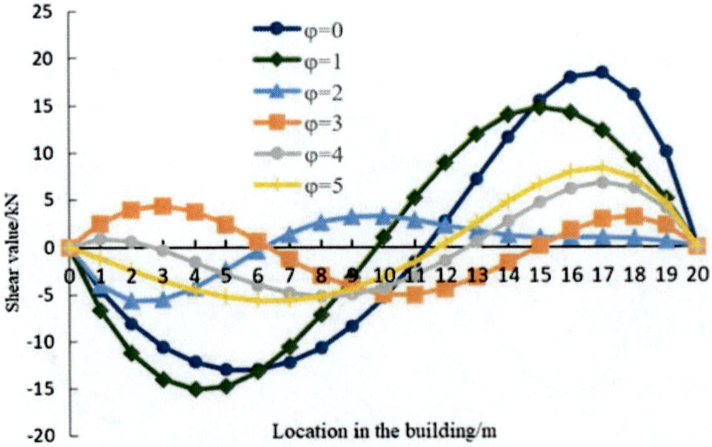

Fig. 6.19 Shear distribution curve of beam (when $S_1 = -20\text{m}$)

6.3 Study on the Impact of Double-Line Shield Tunneling on Adjacent Shallow Foundation Frame Buildings

This section takes ground loss caused by shield construction as the main cause of ground settlement. The mechanical model of the synergistic action of the foundation, foundation and frame structure of the shallow foundation building above the double-line parallel tunnel is derived, and the numerical analysis software 1stOpt is used to solve the problem. The law of the deformation and bending moment of the building in the shield tunneling area is obtained, and the influence of the different tunneling sequence of the double-line parallel shield on the deformatrion of the adjacent building is obtained.

6.3.1 Establishment of Joint Action Mechanical Model

(1) Basic assumptions

It is assumed that the building is parallel to the tunnel axis (i.e., the longitudinal direction of the building is consistent with the tunneling direction), and the transverse horizontal distance between the building and the tunnel axis is y, as shown in Fig. 6.20. The coordinate system as shown in Fig. 6.21 is established: the origin O is established at the ground surface at the left end of the building, and the x axis points to the same direction as the shield tunneling, and the z axis points to the vertical settlement direction of the ground and the building. And $w_1(x)$ is the surface settlement curve, $w(x)$ is the building settlement curve. The distance between the excavation face of shield tunnel and the z axis is s. When the excavation face is on the left side of the z axis, s is positive, indicating that the shield has not reached the

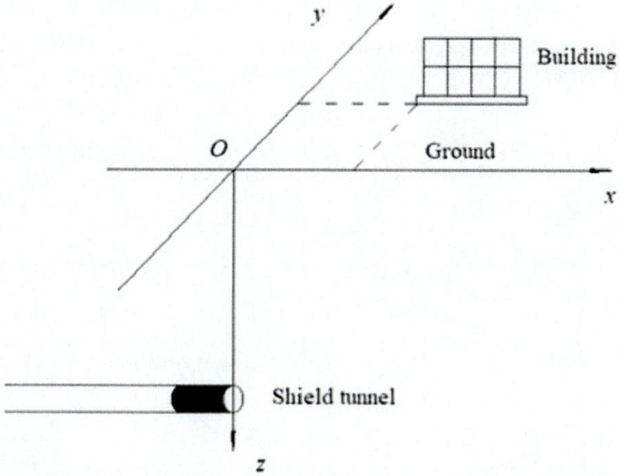

Fig. 6.20 Positional relationship between building and tunnel

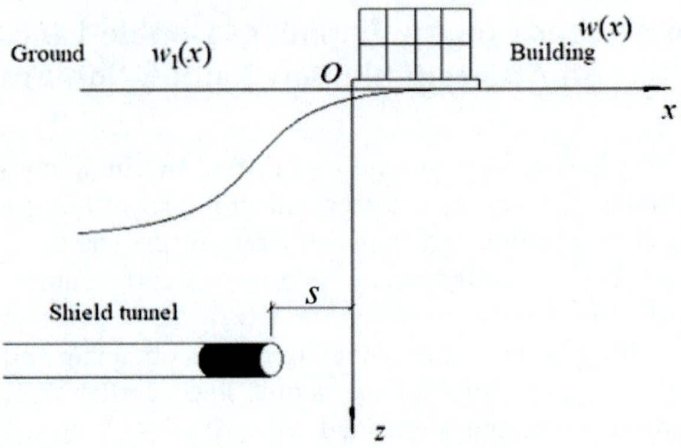

Fig. 6.21 Building and ground subsidence coordinate system

building, and negative when the excavation face is to the right of the axis z, indicating that the shield is passing through or has left the building.

In addition, this section simplifies the foundation—strip foundation—frame structure, and the calculation diagram is shown in Fig. 6.22. Let the length of the building be l, the distance from a certain point on the building to the axis z be i, then the settlement amount of the building at the position of the point be $w(i)$, and the ground settlement amount below the point be $w_1(i)$.

1) Foundation model

According to Winkler elastic foundation theory, the reaction force of building foundation σ_d is proportional to the value of the cut foundation, and it can be obtained as follows:

$$\sigma_d(i) = k[w(i) - w_1(i)] \tag{6.7}$$

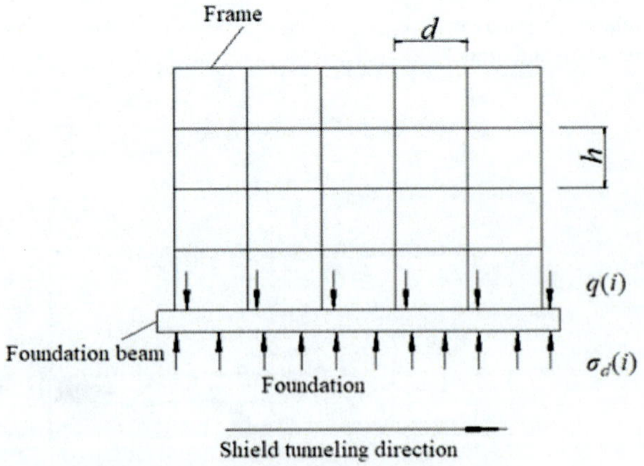

Fig. 6.22 Model calculation diagram

where: $\sigma_d(i)$ is the foundation reaction received at any point on the foundation at the bottom of the building, the unit is kN/m^2, k is the foundation bed coefficient, the unit is kN/m^3.

2) The interaction model of strip foundation and frame structure

The strip foundation building is simplified into a shear beam on an elastic foundation constrained by the superstructure. The differential equation of the flexural deflection of the shear beam is as follows:

$$EJ\frac{d^4w(i)}{di^4} - (GF + g)\frac{d^2w(i)}{di^2} = q - \sigma_d(i) \tag{6.8}$$

where, *EJ* is the bending stiffness of the foundation beam, *GF* is the vertical shear stiffness of the frame, *g* is the constraint line stiffness of the bottom column end, and *q* is the vertical load of the building on the foundation, the unit is kN/m^2).

3) Ground deformation model

See point 3 in Sects. 6.2.1.

(2) Synergy model

Equations (6.4) and (6.7) are substituted into Eq. (6.8) to obtain the flexural differential equation of shear beam on elastic foundation:

$$\begin{aligned} EJ\frac{d^4w(i)}{di^4} - (GF + g)\frac{d^2w(i)}{di^2} + kw(i) = q &+ \frac{kh_1a_1^2}{2(y_1^2 + h_1^2)}\left[1 - \frac{i + s_1}{\sqrt{(i + s_1)^2 + y_1^2 + h_1^2}}\right] \\ &+ \frac{kh_2a_2^2}{2(y_2^2 + h_2^2)}\left[1 - \frac{i + s_2}{\sqrt{(i + s_2)^2 + y_2^2 + h_2^2}}\right] \end{aligned} \tag{6.9}$$

Equation (6.9) is the differential equation of synergistic action of the building foundation, strip foundation and frame structure above the double-line parallel shield tunnel.

(3) Others

Refer to Sect. 4.2.1 for boundary conditions, solution process of differential equation and calculation of building bending moment.

6.3.2 Calculation Conditions

It is assumed that the driving direction of the double-line parallel tunnel is consistent with the longitudinal direction of the building. The axis of the double-line parallel

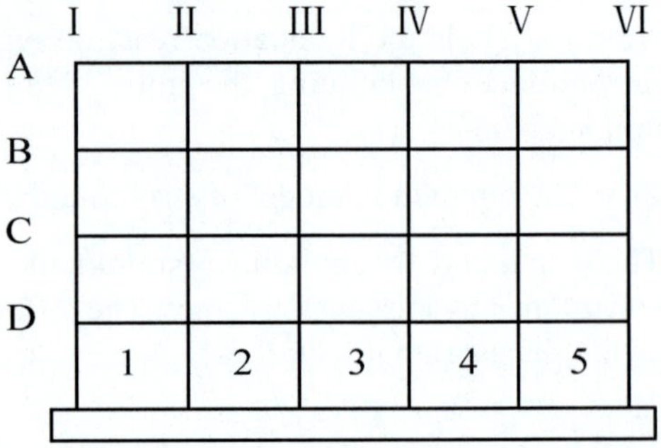

Fig. 6.23 Schematic diagram of a framed building

tunnel is buried at a depth of 12 m and the center distance of the circle is 12 m. The shield is 7.2 m long and 6.34 m in diameter. The lining has an outer diameter of 6.2 m, an inner diameter of 5.5 m, a ring width of 1.2 m and a thickness of 0.35 m. C50 concrete is used for casting. The building is located in the middle of the left and right tunnel (that is $y_1 = y_2 = 6m$). It is a reinforced concrete frame structure with C30 concrete pouring. There are 4 floors above the ground, each height is 3.6 m (including plate thickness), and the spacing is 4 m in the direction of shield tunneling. The column size is 400 mm × 400 mm, the beam size is 300 mm × 550 mm, and the floor thickness is 100 mm. The foundation is strip foundation with a section width of 800 mm and a height of 1000 mm. The elastic modulus of reinforced concrete was set at 30,000 MPa, the foundation bed coefficient was set at 10000 kN/m^3, the vertical load of the building on the foundation was set at 100 kN/m^2, and the soil loss rate was 1% during construction.

Named framework structure horizontal each bay 1–5, each column named I–VI, vertical floor named A–D, as shown in Fig. 6.23.

6.3.3 Calculation and Analysis of Examples

(1) Drive the left and right tunnels simultaneously ($s_1 = s_2$)

Take s_1 as 40, 25, 10, 0, −10, −20, −30, −40 and −50 respectively to simulate the process of the shield reaching, passing through and leaving the building. Each parameter value is substituted into Eq. (6.9), which is calculated by the numerical calculation software 1stOpt. After sorting out, the building settlement curve, the inclined curve, the curve of bending moment distribution of strip foundation and the curve of bending moment distribution of frame beam are obtained.

In Fig. 6.24, the positive value of the vertical axis of the settlement curve is settlement. It can be seen that the frame structure is inclined to one side as a whole when the shield excavation face is 40 m from the building until the excavation face reaches the left end of the building. In the process of the excavation face crossing the building, the settlement of each point on the building keeps changing with the

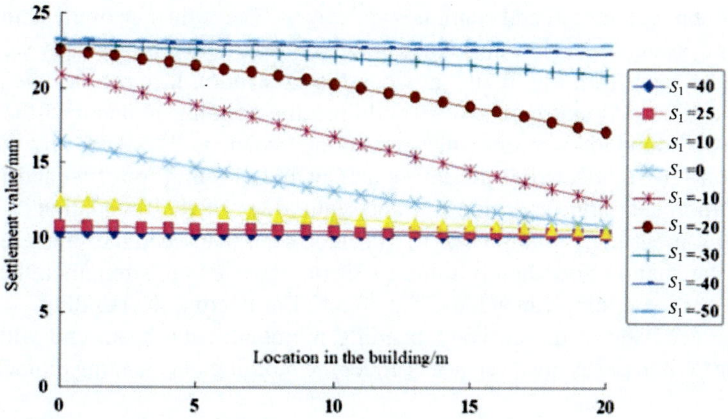

Fig. 6.24 Building settlement distribution curve ($s_1 = s_2$)

excavation face. In the process of excavation face leaving and staying away from the building, the settlement of the right part of the building is large, which makes the building stand upright again, and the maximum settlement value increases obviously, which should be paid attention to in the construction.

In addition, the settlement deformation in the building is basically consistent with the surface deformation at the corresponding position: as the excavation approaches the building, the subsidence of the left end of the building will gradually increase and accelerate, while the subsidence of the right end will change little. It can be seen from Fig. 6.25 that the tilt of the building is the largest when the excavation face is directly below it, and the difference settlement between the left and right ends of the building is the largest, reaching 8.5 mm. When the shield tunnelling surface leaves the building, the settlement at the left end of the building reaches the maximum and

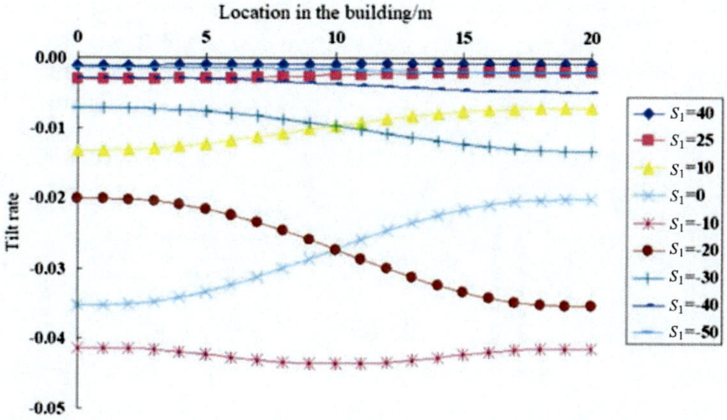

Fig. 6.25 Building slope distribution curve ($s_1 = s_2$)

the settlement at the right end continues to develop. The influence of the tunneling of the backing structure on the subsidence of the frame structure is gradually weakened.

It can be seen from Fig. 6.26 that when the excavation face reaches the position below the left and right ends of the building, the bending moment in the middle of the strip foundation reaches the maximum, which is 27.3 kN · m. When the excavation face is located at the lower half of the building, positive and negative bending moments appear on the left and right sides of the foundation, while the bending moment in the middle is zero. It can be seen that when the shield tunneling through the shallow foundation frame structure, measures to strengthen the central stiffness of the building can achieve the effect of protecting the building.

Figure 6.27 shows the curve of bending moment at the beam end with shield tunneling. When the excavation face is under the building, the bending moment of the

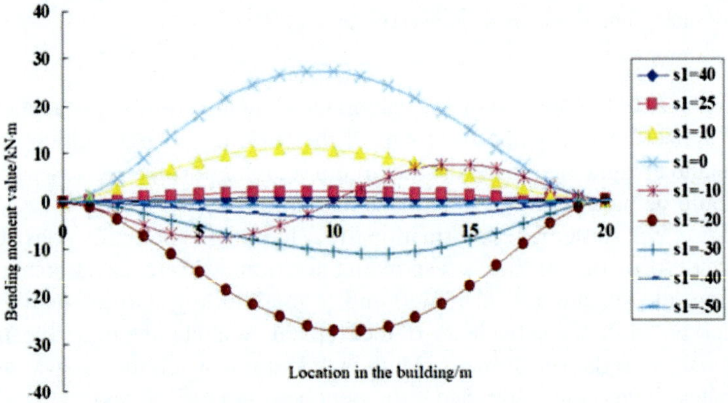

Fig. 6.26 Curve of moment distribution of strip foundation ($s_1 = s_2$)

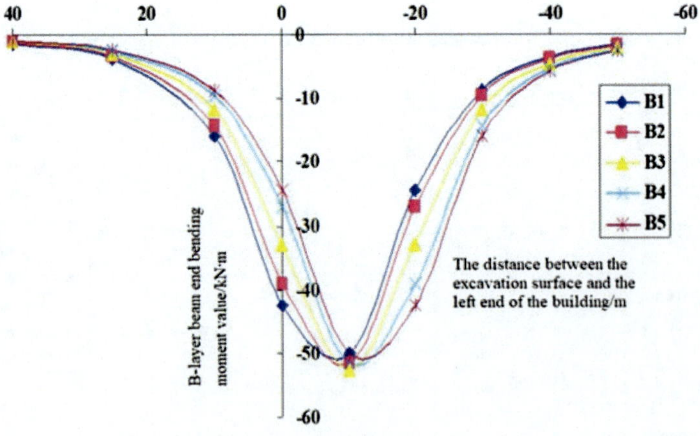

Fig. 6.27 Bending moment change curves of each beam end in Layer B ($s_1 = s_2$)

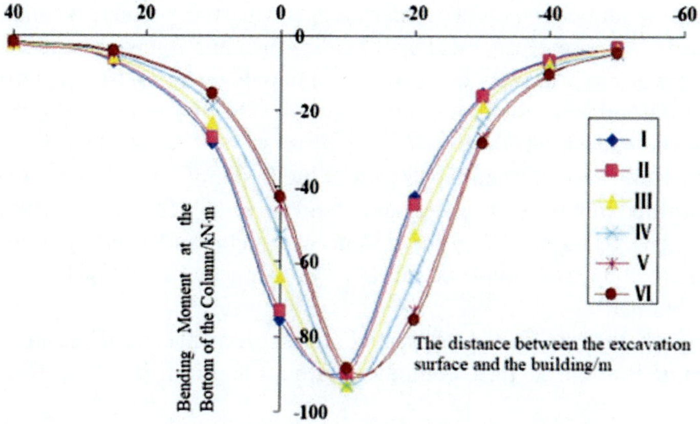

Fig. 6.28 Curve of restrained bending moment at the bottom of column ($s_1 = s_2$)

frame structure of the building changes greatly. It can be seen that the bending moment gradually increases with the approach of the excavation face. When the excavation face reaches the vicinity directly below the beam, the bending moment reaches its maximum value of 52.7 kN · m. After that, the bending moment gradually decreases, and the entire curve presents an approximately symmetric V-shaped distribution. The maximum bending moment of the B-layer beam is greater than that of the foundation beam, indicating that the stress of the frame beam in the driving area is greater than that of the shallow foundation beam, which needs to be paid attention to in the construction and design.

The variation law of the restrained moment of the bottom column on the foundation as shown in Fig. 6.28 is consistent with the bending moment variation diagram of the frame beam, but the maximum bending moment value is larger, indicating that the connection between the column and the foundation is also more vulnerable to damage during tunneling. In addition, literature [15] shows that the bending moment of the beam and column of the upper frame structure of the building is mainly determined by the tilt rate of the building. The greater the tilt rate is, the greater the constrained bending moment of the beam end and the bottom column on the foundation will be naturally. Therefore, in the construction process, it is necessary to strengthen the monitoring of settlement difference and tilt rate at the end of the building, so as to better control the impact of shield tunneling on the bending moment of the adjacent shallow foundation frame building.

(2) Left and right tunnel driving successively ($s_1 - s_2 = -30$)

In practical engineering construction, the left and right lines of the double-line parallel tunnel are not tunneling simultaneously. In order to reduce the influence of the later tunneling shield on the first tunneling shield, the specification requires that the spacing between the two shields should be more than 100 m apart. However, in some cases, the distance between the backward tunneling shield and the first tunneling shield is very small, just like the special case in which the backward tunneling shield exceeds the first tunneling shield in a certain interval of Shanghai's orbital communication Line 7 [19]. Therefore, this paper studies the situation where the left and right excavation faces are relatively close apart, and analyzes the settlement and bending moment variation law of the upper building under this working condition. Here, the distance of the excavation face is 30 m, that is, $s_1 - s_2 = -30$ (s_1 is the distance between the excavation face and the z axis of the first tunneling tunnel). Each parameter is substituted into the synergy equation and solved by 1stOpt. After finishing, the settlement curve of the building, the tilt curve, the curve of bending moment distribution of strip foundation and the curve of bending moment distribution of frame beam are obtained.

From Figs. 6.29 and 6.30 shows parallel final settlement caused by shield tunneling building value is about 23 mm, compared with around in front of the line and at the

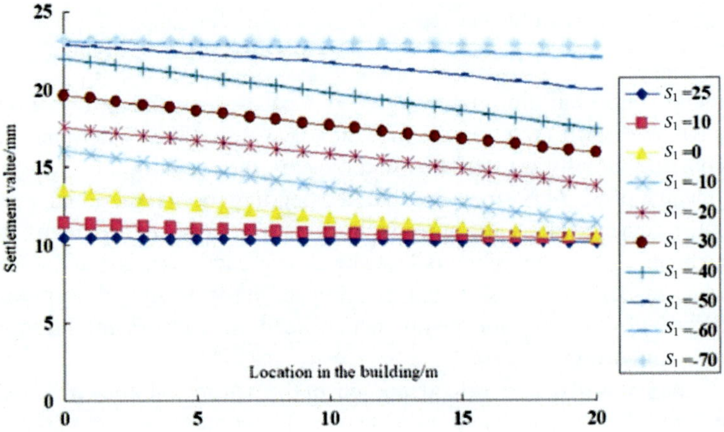

Fig. 6.29 Building settlement distribution curve ($s_1 - s_2 = -30$)

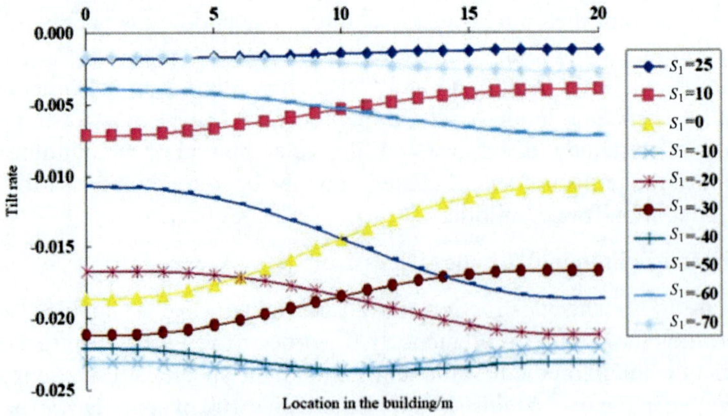

Fig. 6.30 Building slope distribution curve ($s_1 - s_2 = -30$)

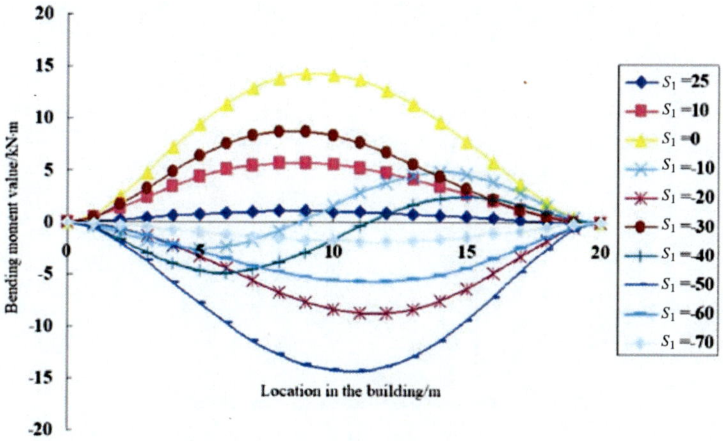

Fig. 6.31 Curve of moment distribution of strip foundation ($s_1 - s_2 = -30$)

same time driving conditions is the same, a gradual, but the change trend of building subsidence and tilt rate is smaller, largest and biggest tilt rate appeared after the tunneling of shield and a prior tunneling shield excavation face directly the building, which we should pay attention to in the construction.

Figure 6.31 knowable by foundation beam bending moment distribution become more complicated, in shield tunneling through successively has experienced several times larger changes in the process, and the influence range is larger, first tunneling shield does not leave the building 50 to 25 m to the distance, the entire process influenced by shield tunneling, the maximum bending moment of foundation beams of the position remains unchanged, the size of 14.2 kN · m., but its value is much smaller than double driving conditions at the same time.

By a Figs. 6.32 and 6.33 the frame structure of the bending moment with shield tunneling approximate symmetrical W type distribution, and starting with tunneling shield excavation plane arrived to leave the building process of 30 m, in a larger value, maximum value appeared on the biggest building tilting, B layer beam end bending moment and the column bottom constraint maximum bending moment are respectively 28.6, 25.3 kN · m. Moreover, the bending moment of the frame structure and the foundation beam of the building will vary with the distance between the two excavation faces. The farther the distance is, the smaller the maximum bending moment will be. However, when the distance of the excavation face exceeds a certain value, the maximum bending moment of the building will remain basically unchanged. Moreover, the maximum bending moment of the building beam and column caused by successive tunneling is obviously smaller than that of the simultaneous tunneling condition, but the maximum bending moment range of the building is larger than that of the simultaneous tunneling condition, which is worth noting in construction and design.

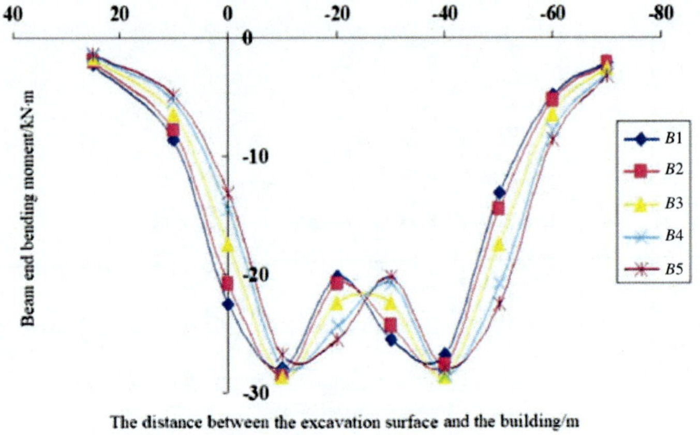

Fig. 6.32 Curve of bending moment at the beam end of Layer B ($s_1 - s_2 = -30$)

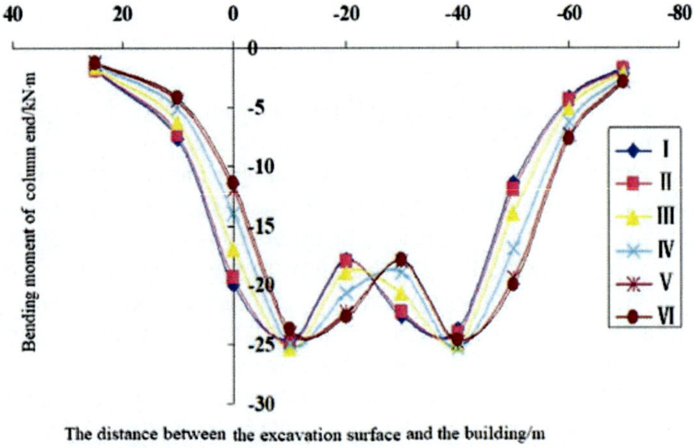

Fig. 6.33 Bending moment change curve at column end ($s_1 - s_2 = -30$)

6.4 Study on the Impact of Double Shield Tunneling on Adjacent Shallow Foundation Frame Buildings

In this section, the influence of the construction of double-circular shield tunnel on the adjacent shallow foundation frame buildings is studied by referring to the mining area synergistic model method, the mechanical model and analytical solution of the synergistic action of foundation, foundation and frame structures of strip foundation buildings are derived, and the longitudinal deformation and internal force variation rules of buildings in tunneling area are studied.

6.4.1 Establishment of Joint Action Mechanical Model

(1) Basic assumptions

Two coordinate systems are established as shown in Fig. 6.34. The surface subsidence coordinate system is: $w_1(j) - O_1 - j$, the origin O_1 is established at the surface above the excavation face; the horizontal j axis points to the same direction as the double shield tunnelling machine; the vertical coordinate $w_1(j)$ is the surface subsidence at point j. The building subsidence coordinate system is $w(x) - O - x$, the origin O_1 is established at the left end of the building's surface, the distance is j (j is positive before the excavation face reaches the building, and negative after the excavation face passes through the building and leaves), the horizontal coordinate x axis is consistent with the j axis, the subsidence of any point x in the building is $w(x)$, and the corresponding surface point subsidence is $w_1(j + x)$.

The foundation—strip foundation—frame structure above the axis of the double shield tunnel is simplified, and the strip foundation—frame building is simplified into a shear-bending beam on the elastic foundation, as shown in Fig. 6.22. It is assumed that the deformation of the superstructure is caused by ground settlement during tunneling, and the possibility of stripping between shallow foundation and foundation is not considered in this section.

1) Foundation model

See (1) in Sect. 6.2.1.

2) The interaction model of strip foundation and frame structure

See (2) in Sect. 6.3.1.

3) Ground deformation model

The calculation models of soil displacement caused by double-circle shield construction mainly include the equivalent great circle model, the double-circle superposition

Fig. 6.34 Building and ground subsidence coordinate system

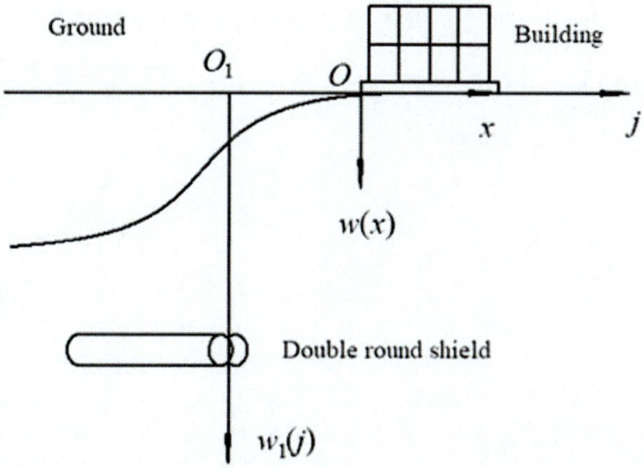

model and the non-uniform convergence model of random medium. The double-circle superposition model is still the most widely used. The calculated results are closer to the measured values than the equivalent great circle model, and the error is only about 8.5% compared with the random medium non-uniform convergence model, and the practical application is more convenient. Moreover, the research in literature [13] shows that soil loss is the main influencing factor of soil displacement in the construction of double-circular shield. In this section, considering the complexity of the model, the surface settlement caused by the construction of the double shield is caused by soil loss, and can be seen as the superposition of the land settlement caused by the separate action of two intersecting single shield [20], as shown in Fig. 6.35.

According to the calculation theory of Sagaseta, the calculation formula of longitudinal ground settlement above the axis of the double round tunnel is

$$w_1(x) = \frac{a^2 h}{(L/2)^2 + h^2}\left[1 - \frac{x}{\sqrt{x^2 + (L/2)^2 + h^2}}\right] \tag{6.10}$$

where, x is the horizontal distance from the digging surface in the driving direction of the shield; y is the transverse horizontal distance from the tunnel axis; a is the loss radius of soil point; h is the depth of the tunnel, the same as below.

(2) Synergy model

Equations (6.1) and (6.10) are substituted into Eq. (6.8) to obtain the flexural differential equation of shear beam on elastic foundation

$$\begin{aligned} EJ\frac{d^4 w(x)}{dx^4} - (GF + g)\frac{d^2 w(x)}{dx^2} + kw(x) = q \\ + \frac{kha^2}{(L/2)^2 + h^2}\left[1 - \frac{j + x}{\sqrt{(j + x)^2 + (L/2)^2 + h^2}}\right] \end{aligned} \tag{6.11}$$

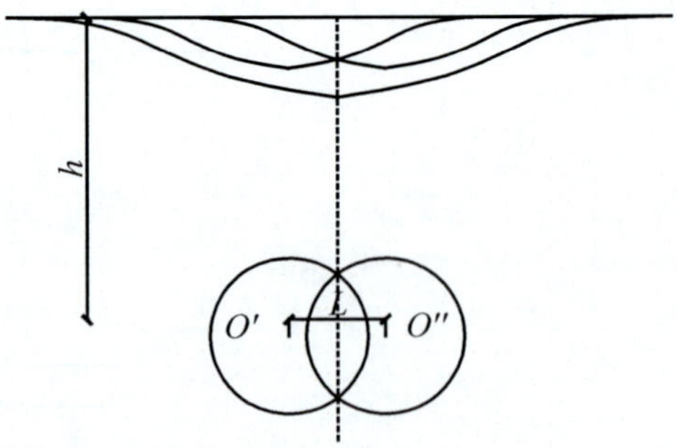

Fig. 6.35 Schematic diagram of double circle superposition model

Equation (6.11) is the differential equation of the combined action of the building foundation—shallow foundation—frame structure above the axis of the double-circular shield tunnel.

(3) Others

Refer to Sect. 4.2.1 for boundary conditions, solution process of differential equation and calculation of building bending moment.

6.4.2 Calculation Conditions

Furthermore, the deformation and stress analysis of the buildings above the tunneling area are carried out. Calculation basic conditions: the building is reinforced concrete frame structure, using C_{30} concrete pouring. There are 4 floors above ground, each height is 3.6 m (including plate thickness), and the spacing is 4 m along the direction of shield tunneling. The column size is 300 mm × 300 mm, the beam size is 300 mm × 550 mm and the floor thickness is 100 mm. The foundation is strip foundation and the cross section size is 1000 mm × 800 mm. The elastic modulus of reinforced concrete is set at 30,000 MPa and the coefficient of foundation bed at 5000 kN/m^3, the vertical load of the building is 200 kN/m^2. Double circle shield machine cutting face type is glasses, section size is Φ6.52 m × W11.12 m (diameter × width), double center distance 4600 mm, the tunnel axis depth of 9.1 m, the soil loss rate of 2%.

In order to clear, transverse each bay is named the building 1, 2, 3, 4, 5, each column named I, II, III, IV, V, VI longitudinal floor named A, B, C, D, as shown in Fig. 6.23.

6.4.3 Case Calculation and Analysis

(1) Analysis of building deformation and internal force change

By using 1stopt software for programming calculation, the subsidence curve, the change of beam, column and foundation internal force of the building in the tunneling area of the double round shield is obtained. As shown in Fig. 6.36, when the double-circle shield machine does not reach the building, that is, at $j > 20$ m, the building has settled to a certain extent due to the load of the structure itself, and the settlement is related to the foundation property and the building load, etc., which should be paid attention to in the construction of double-circle shield. When the shield machine approaches the building gradually, the settlement of the building tilts to one side as a whole, which is consistent with the finite element simulation results in literature [21]. Moreover, the subsidence value of the building varies uniformly from the left end to the right end, and the settlement starts at a distance close to the shield machine. When the excavation surface of the shield is 10 m below the building, the settlement

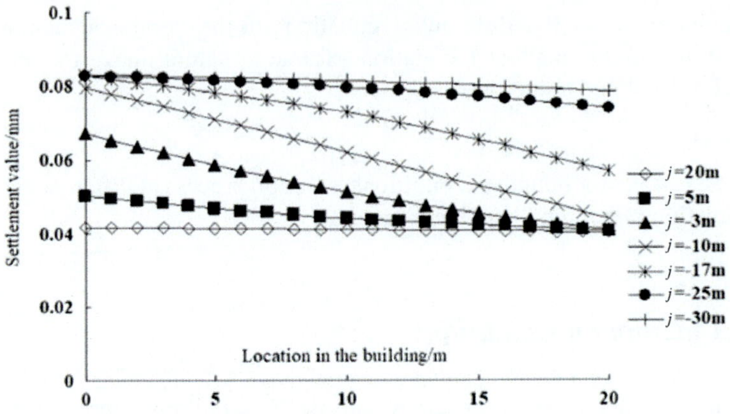

Fig. 6.36 Deformation diagram of frame building

difference between the head and the tail of the building is the largest. In this case, the additional settlement of the building is equivalent to its own settlement value, which indicates that the construction of the double shield has caused the larger secondary deformation of the building.

The additional bending moment and shear force of any section on the shallow foundation of the building vary greatly in the process of shield tunneling, as shown in Fig. 6.37a and b. When shield tunneling lead to displacement of surrounding soil, the foundation of the additional bending moment and shear force will also begin to produce, when the excavation surface to about 1/5, 4/5 position directly (maximum negative and positive curvature buildings located on the surface subsidence curve), base in central bending moment value maximum, 103.34 kN · m, the distribution of the bending moment is approximately symmetrical concave or convex curve; When the excavation face is located between the two, positive and negative bending moments occur simultaneously on the foundation of the building. When excavation surface in the middle of the building, the maximum positive and negative bending moment, same as ±32.50 kN · m, respectively in 3/4, 1/4 location near buildings, the distribution of bending moment is antisymmetric, and length of the positive and negative bending moment for buildings, and the range of influence in the excavation of shield machine is relatively small (about 50 m), causing buildings left sides of the tilt is different, the positive and negative curvature; Finally, when the settlement of the building is stable, the value of additional bending moment of the foundation is also close to zero.

No matter where the excavation face is located, positive and negative shear forces always exist on the foundation, which is also due to the initial settlement of the building itself. When the excavation face reaches about 1/5 of the building, the maximum positive shear force is generated near 1/4 of the foundation, which is 16.93 kN; the maximum negative shear force is generated near 3/4 of the foundation, which is −14.95 kN; the central shear force is zero, and the shear distribution is

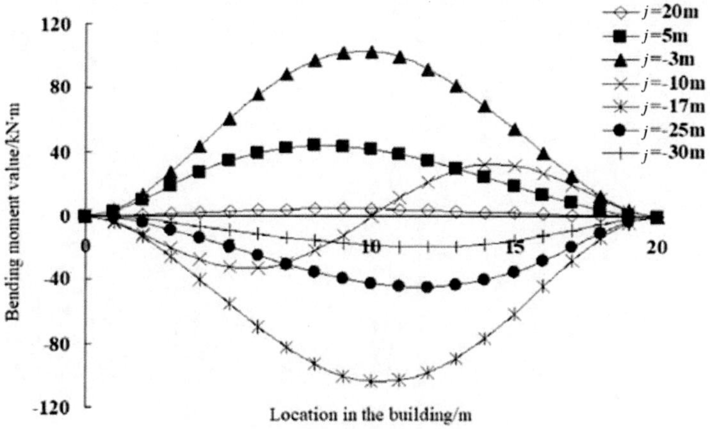

(a) Additional bending moment of strip foundation

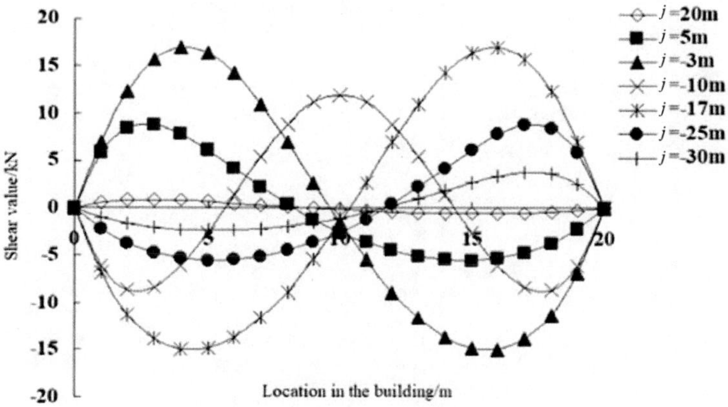

(b) Additional shear on strip foundation

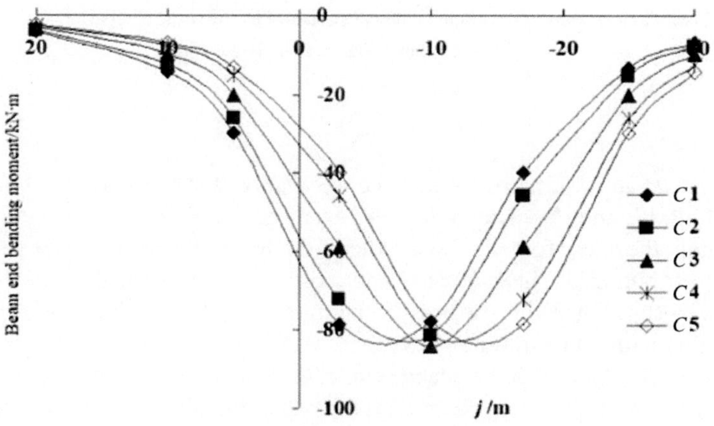

(c) C beam end bending moment

Fig. 6.37 Variation of internal forces in frame buildings

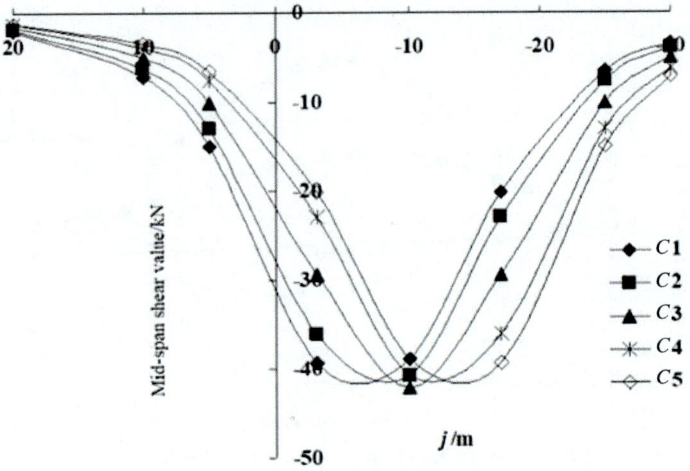

(d) Midspan shear of dc beam

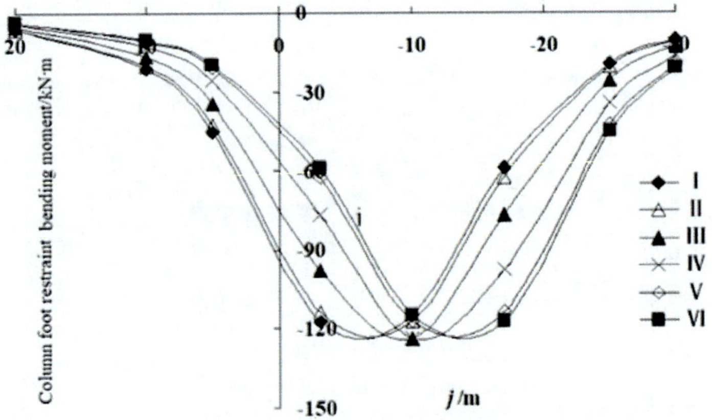

(e) The foot constraint bending moment generated by the constraint between the bottom column and the strip foundation

Fig. 6.37 (continued)

approximately antisymmetric. When the excavation surface reaches about 4/5 of the building, it's the opposite; When the excavated surface reaches the middle of the building, the shear force in the middle of the foundation reaches the maximum, and the distribution of shear force is approximately symmetrical. This shows that measures to strengthen the stiffness in the middle of the building can achieve the effect of protecting the building.

Figure 6.37c and d show the change curves of mid-span shear force and beam end bending moment of the beam along with the tunneling of the double circular shield. It can be seen that the shear force and bending moment of the beams on each floor of the frame building increase gradually with the approach of the excavation face, and the maximum internal force value of the beams on each floor of C is not the same. When the excavated surface of the shield machine reaches the vicinity of the right under the beam, the shear force and bending moment reach the maximum value, and then the bending moment and shear force gradually decrease, and the whole curve is approximately symmetric. The maximum bending moment at the beam end of Layer C is close to that of the foundation beam, but the maximum shear force in the middle span is greater than that of the foundation beam, indicating that the stress on the frame beam in the driving zone is greater than that on the shallow foundation beam, which should be paid attention to in the construction and design.

Figure 6.37e shows that the constraint moment law of the bottom column on the foundation is consistent with the diagram of internal force variation of the frame beam, but the maximum bending moment value is larger, and the constraint moment of each column on the foundation is greater than that of the beam end of the C-layer beam, which indicates that the connection between the column and the foundation is also more vulnerable to damage during tunneling. In addition, the internal force of the beam-column of the upper frame structure of the building is mainly determined by the tilt of the building. The larger the tilt rate is, the larger the mid-span shear force of the beam, the bending moment of the beam end and the constrained bending moment of the bottom column on the foundation will be naturally. Therefore, in the construction process of double circular shield, it is necessary to strengthen the monitoring of settlement difference and tilt rate at the end and head of the building, so as to better control the influence of shield tunneling on the internal force of the adjacent shallow foundation frame structure.

(2) The influence of burial depth and soil loss rate on building internal force

The deformation and internal force of buildings in the shield tunneling area are affected by many factors, such as the structure form of buildings, the depth of tunnel and the construction technology of shield tunneling [22]. Referring to existing research results [23], this section only makes a comparative analysis of factors that have a greater impact, such as burial depth and soil loss rate. It can be seen from Fig. 6.37 that the equivalent shear force generated by the constraint of the bottom column on the shallow foundation and the bending moment constrained by the foot of the column are consistent with the change rules of the mid-span shear force of the beam, so the mid-span shear force of the beam is taken as the research object. The beam of floor C in the second branch is taken here for research. Other calculation conditions are as the case of the frame building mentioned above.

1) Influence of tunnel axis burial depth

The depth of tunnel axis is set as 4m, 8m, 12m, and 16 m respectively. Other calculation conditions remain unchanged. The calculation results are shown in Fig. 6.38. When the shield machine is located at the bottom of the building, the mid-span shear force of the beam in Floor C increases as the excavation face approaches. When the excavation face reaches the vicinity of the beam, the shear force reaches the maximum and then decreases gradually. The smaller the buried depth of the tunnel, the greater the maximum shear force, and with the reduction of the buried depth, the

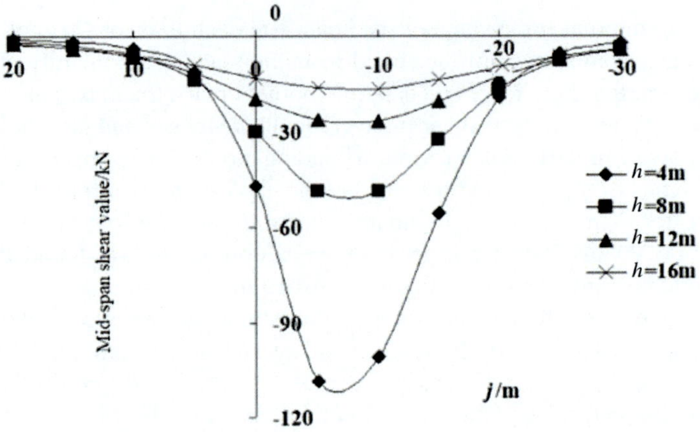

Fig. 6.38 Change curve of shear force across the beam with different buried depth

greater the increase. However, before the shield machine reaches the building, if the distance is more than 10 m, the shear force will be smaller with the lower buried depth. It can be seen that, if the foundation under the building is strengthened, the tunneling scope and buried depth should be considered. If shield tunnel is to pass through important buildings, the effect of protecting buildings can be achieved by reasonably increasing the buried depth of tunnel in design.

2) Influence of soil loss rate

Literature [24] collected a number of domestic shield tunnel construction projects, and analyzed the measured data of soil loss rate. It was found that the soil loss rate of shield tunnel construction in soft soil areas ranged from 0.20 to 3.01%. Accordingly, the soil loss rate in this section is taken as 0.5%, 1%, 1.5%, 2%, 2.5% and 3%, and other calculation conditions remain unchanged. The results are shown in Fig. 6.39. It can be seen that the greater the soil loss rate is, the greater the maximum shear force of the beam in floor C of the building will be. This is because as the soil loss rate increases, the ground settlement increases and the additional internal force of the building also increases. The results show that the maximum shear force in the span of the C-storey beam has a good linear relationship with the soil loss rate.

Therefore, in this section, it is believed that the soil loss rate should be effectively reduced to control the secondary deformation of the building in the tunneling area of the double shield. Moreover, research in literature [25] shows that in soft soil layer, the grouting of shield tail is the key factor to control the stratum settlement caused by the construction of double circular shield. So it is very important to control the grouting at the rear of the double shield reasonably to reduce the deformation of the building. First of all, the grouting behind the tail wall of the double round shield shall be strictly controlled. The initial setting time of the double-liquid grouting shall be controlled within 8–12 s, and the grouting amount shall be guaranteed to be 150–180% of the theoretical gap of the tail of the shield. At the same time, the second

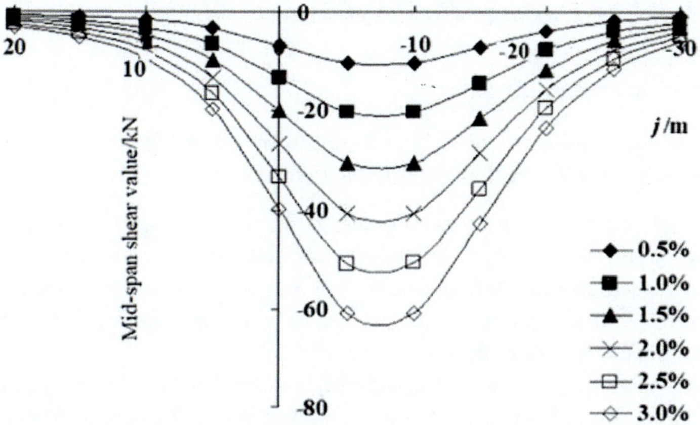

Fig. 6.39 Change curve of mid-span shear force of beam with different soil loss rates

grouting and the follow-up grouting are strengthened to ensure the compaction of the lining pipe segment and the soil gap [26].

6.5 Study on Building Deformation Control Standards

Shield tunnel construction leads to the deformation of surrounding soil and causes the adjacent buildings to move and deform, resulting in additional stresses. These additional stresses are an important factor for the deformation and failure of surface buildings. Due to these inevitable adverse factors, the project can only make better use of the monitoring system and cooperate with the simultaneous repair and reinforcement work to remedy, especially for some important buildings on the shield tunnel construction line, is the key object of protection. The synergy of this section by Sect. 6.2 mechanical model as the research object, the reference, the measured deformation due to shield tunneling value, combined with the settlement of the ground buildings for safety standards, puts forward the appropriate evaluation of building safety evaluation theory, for the future similar to the double parallel shield tunnel adjacent shallow foundation building construction project to provide the reference.

6.5.1 *Building Damage Risk Assessment*

(1) Definition of building damage

According to the Skempton etc. [27] and studies by other scholars generally classify building damage into the following three categories:

1) Structural damage. Structural damage is mainly damage to the appearance of components, such as cracks in wall panels, floors and building surfaces. Cracks >0.5 mm wide in painted walls and 1.0 mm wide in masonry walls and gross concrete walls are generally considered to be the maximum size of cracks that can be observed by building occupants.
2) Functional damage. Functional damage is mainly caused by structural or structural functional obstacles, such as doors and Windows cannot be opened, wall or floor tilt, gas or water pipe bending and rupture, decorative surface cracking and peeling, etc. Functional damage generally does not require structural repair.
3) Structural damage. Structural damage often affects the stability of the structure. Such damage includes cracking and serious deformation of the main stress components of the building, such as beams, columns, floors and load-bearing walls. Burland [28] On the basis of previous researches, according to the degree of difficulty in repairing the maximum crack of masonry wall, a classification standard of building damage level is given.

Table 6.1 details the classification levels of the various damage levels and their descriptions [29].

Among them:

1) The table 6.1 evaluates the level of building damage based on the difficulty of crack repair.
2) Crack width is not the only criterion for assessment. The location and number of cracks should be taken into account.
3) A local horizontal or vertical deviation of more than 1/100 will clearly be observed, and an overall deviation of more than 1/150 will cause visual discomfort.

(2) Risk assessment of building damage

The above model is also used for calculation and analysis. By referring to [29–31], the following three standards are given for the determination of building safety risk grade, and checking calculation is carried out according to different standards to obtain different levels of building analysis damage. The control quantity setting of the control standard is different. For example, some consider the existing settlement of the building itself and set the control standard as the total amount of the existing settlement, while some only consider the increment of the settlement of the building.

At the same time, two kinds of standards are adopted for settlement control, and the most conservative and safe one is to judge the dangerous state of the building when it reaches the warning value. But in the end, it is necessary to analyze the most

Table 6.1 Damage degree and description table

Category	Damage	Damage description (underlined section indicates how easy it is to repair)
0	Can be ignored	Capillary crack, crack width less than 0.1 mm
1	Very mild	Fine cracks can be dealt with easily through general decoration. Damage is generally limited to finishes on the interior walls. Cracks in brickwork or mortar can be viewed up close. Typical crack width up to 1 mm
2	A slight	Cracks can be easily filled in and may need to be redecorated. Cracks that often occur can be concealed by a suitable lining. There are obvious cracks on the exterior surface of the building and the joints should be pointed to prevent leakage through air. Door and window openings are slightly affected. Typical crack width up to 5 mm
3	Moderate	The crack must be mended. The outer brick wall needs to be re-pointed, and a small part of it may need to be removed and replaced. The doors and Windows are stuck. The pipeline is likely to break. Air permeability and water leakage weakened. The typical crack width can be 5–15 mm, or several cracks wider than 3 mm
4	Serious	Extensive repairs to the building will be required, including the removal or replacement of parts of the walls (especially those above doors and Windows).The door and window frames are twisted and the floor is clearly tilted. Tilting or bulging of the wall, the bearing capacity of the beam is damaged. Pipeline rupture. Typical crack width can be up to 15–25 mm (also related to the number of cracks)
5	Very serious	The building must be partially or completely rebuilt. The beam has lost its bearing capacity and the wall is badly inclined and in need of support. The window was broken with distortion. The structure is in danger of instability. The typical crack width is >25 mm (also related to the number of cracks)

reasonable standards and explanations based on the actual building conditions. In the previous model, in order to obtain reliable analysis data, the driving dynamic process of shield is subdivided and the amount of data is huge. For this section, in order to make simple calculation and get the most obvious conclusion, the maximum settlement in the previous calculation is taken for analysis and study.

1) Incremental calculation of building settlement

Increment of settlement: is the difference value of settlement at a certain point before and after the shield crossing.

2) Calculate the additional slope of the building

The existing slope of the building:

$$k_{ab} = \tan\varphi = \frac{h_a - h_b}{l} \tag{6.12}$$

Slope during shield propulsion:

$$k'_{ab} = \tan \varphi' = \frac{h'_a - h'_b}{l} \tag{6.13}$$

Additional slope:

$$k''_{ab} = \tan \varphi'' = \frac{(h'_a - h_a) - (h'_b - h_b)}{l} \tag{6.14}$$

According to the additional slope formula, the longer the longitudinal direction of the building, or the higher the overall stiffness of the building, the less the threat of damage caused by the tilt of the building.

If the building itself has certain tilt, and shield through the direction is to tilt has a role in intensifying, that could reach a state in the process of shield tunnel crossing point, building up to 0.2% or more large tilt rate, will cause varying degrees of damage of the building, at this point it is necessary for building protection reinforcement measures on the ground. On the contrary, if the inclination of the surface building caused by shield construction is mutually offset with the original inclination of the building, then the settlement problem brought by the construction will weaken the dangerous state of the original building within a certain range, which is relatively beneficial to the safety of the building. However, if the tilt problem caused by the construction is too serious, it will cause a reverse tilt, which may still pose a threat to the building.

6.5.2 Building Settlement Control Standard

(1) Building damage risk assessment based on maximum slope and typical value of settlement

Reference [29] standard carries out risk assessment for typical settlement value, maximum slope and damage degree of general buildings, as shown in Table 6.2.

In this model, simultaneous tunneling of double-line shield is the fastest and the most violent, so the simultaneous tunneling is taken as the calculation condition. Under this working condition, the tilt of the original building is close to 0, so the original tilt is not considered. The original settlement of each point is basically −40.86 mm, and the final settlement of each point is basically −82.8 mm. The calculation shows that the maximum settlement difference is 41.24 mm, and the maximum tilt reaches 0.202% in the whole tunneling process.

According to the above table, it can be seen that there is a certain settlement of the building itself before construction, and the building does not tilt. After construction, the settlement difference is up to 41.24 mm, and the maximum tilt of the building is up to 0.202%, which is risk level 2, with slight risk. Therefore, for the safety of buildings and the protection of historic buildings, the building itself should be done before the construction of reinforcement and repair protection work.

Table 6.2 Assessment of building damage risk with maximum slope and maximum settlement

Risk level	Maximum building slope	Maximum building settlement/mm	Risk description
1	<1/500	<10	Negligible risk: surface damage is unlikely
2	1/500–1/200	10–50	Minor risks: possible surface damage that does not affect important structures
3	1/200–1/50	50–75	Medium risk: expected surface damage and possible structural damage may result in rigid piping damage
4	>1/50	>75	High risk: anticipated structural damage to the building. Failure of rigid piping, possible failure of other piping

(2) The control standard of the building according to the deformation amount and the overall rotational momentum of the building

Burland and Wroth gave definitions of various deformation variables of buildings, which have been widely recognized by relevant studies, as shown in Fig. 6.40 [29].

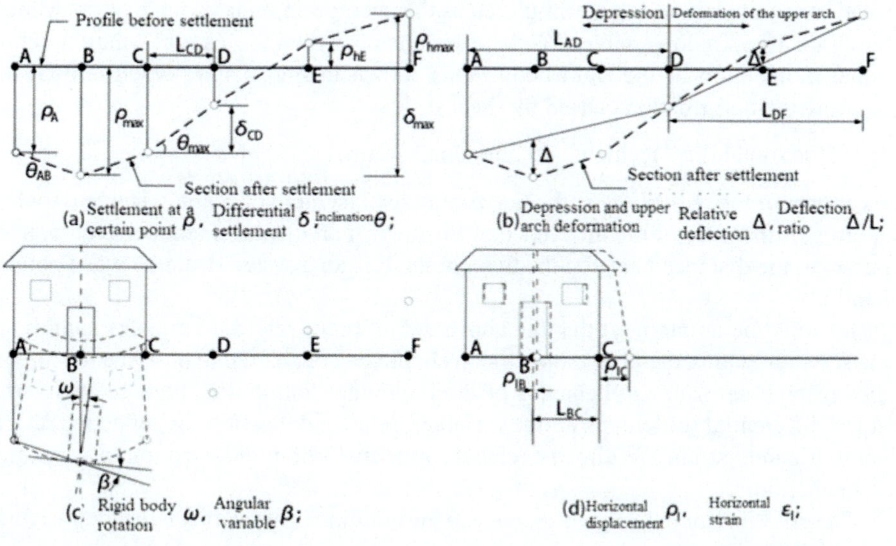

Fig. 6.40 Schematic diagram of building parameters

1) Settlement, differential settlement and Angle of rotation

Figure 6.40a shows ρ_i, the downward displacement at point *i*, i.e., the settlement value; ρ_{hi} is the upward displacement of the first point *i*, that is, the upward lift value. δ_{ij} is the difference settlement between the first point *i* and the first point *j*. The Angle θ_{ij} is the ratio of the difference settlement δ_{ij} between the first point *I* and the first point *j* and the distance L_{ij} between these two points, which is used to describe the slope of the settlement curve.

2) Depression deformation, upper arch deformation, relative deflection and deflection ratio

As shown in Fig. 6.40b, the deformation of buildings can be divided into two modes: sag and upper arch, where sag means concave on the subsidence profile curve of buildings, while upper arch means concave on the subsidence profile curve of buildings. Point D in the figure is the boundary point of sag and upper arch deformation. Relative deflection Δ is building subsidence profile curve and the maximum distance between the two reference point of attachment. Deflection ratio of relative deflection Δ and the ratio of the distance between the two reference points, that is, Δ/L. The deflection ratio can be used to approximate the curvature of the settlement curve and is generally related to the deformation caused by bending.

3) Rotational momentum and angular variables of the rigid body

As shown in Fig. 6.40c, the rotational momentum of the rigid body of the whole structure is represented by ω. The rigid body rotation of the building will not cause the distortion of the building components, so the beam, column, wall and foundation of the building will not crack. In calculation, the value of the rigid body rotation momentum ω can be simplified as the average angular variable of the whole building. The Angle variable β is the difference between the Angle θ and the rotational momentum of the rigid body ω, as shown in Fig. 6.40a, which is used to measure the deformation caused by shear.

4) Horizontal displacement and horizontal strain

As shown in Fig. 6.40d, ρ_{li} is the horizontal displacement of point *i*. The horizontal strain ε_l is the ratio of the difference of the horizontal displacement between points *i* and *j* to the distance between the two points. It is an average strain between points *i* and *j*.

It should be pointed out that the above definition of relevant variables applies to the situation within the plane, and torsion should be considered to describe the three-dimensional deformation behavior of the building. Among the above related variables, differential settlement, Angle variables, relative deflection (or deflection ratio) and horizontal strain are directly related to the distortion, deformation or cracking of the building.

The criterion for judging the damage to the building is based on the Angle variable β, which can be obtained as follows:

$$\beta_{ij} = \theta_{ij} - \omega = \frac{\delta_{ij}}{\mathrm{L}_{ij}} - \omega \tag{6.15}$$

where, ω is the rotational momentum of the rigid body of the building, $\omega = \frac{\sum \theta_{ij}}{L}$.

The settlement of the model is within 0–20 m of the building. Every 1 m is a calculation point. The 0–20 m of the building is divided into 19 intervals according to the 1 m spacing S_1.

1) If the interval is 0, i.e., the interval of 0–1 m in the building, trial calculation: If under the condition of synchronous driving, the angular variable of the interval of 0–1 m in the building. $S_1 = -10$ m is calculated:

$$\theta = 0.00136/1 = 0.00136, \omega = \frac{\sum \theta}{L} = \frac{0.019656}{20} = 0.0009828$$

$$\beta = \theta - \frac{\sum \theta}{L} 00.00136$$

It can be seen that the repeated calculation of Angle variables is a heavy workload. In order to achieve the purpose of analysis, after completing the calculation of Angle variables, we took the maximum value in each case as the representative value to analyze the failure characteristics of buildings.

Maximum Angle variable:

$$\beta_{\max} = \theta_{\max} - \omega \tag{6.16}$$

2) Calculation and analysis of angular variables

Take the shield spacing as 0, $1D$, $3D$ and $5D$ as the calculation objects, and take the distance S_1 as 0m, −5m, −10m, −15m and −20 m. The calculation results are shown in Figs. 6.41, 6.42, 6.43, 6.44, 6.45 and 6.46.

Excavated plane just arrived in the left side and leave the building right end two periods, shield tunnel excavation face are calculated point at the top of the building the maximum Angle variable, there has been a marked that the building of small area, such as shield machine arrived ($S_1 = 0$), within the range of 0 (building within 0–1 m) tilt in the most severe, Angle variable, to the right shows that leaning to the right between each area is more and more small, within the range of 10 (building 10–11 m) close to 0, the buildings here smoothly, right inverse tilt state. When the shield shield reached a quarter of the building ($S_1 = -5$ m) and the Angle variable was still close to the initial arrival, the local area tilt of the building changed seriously, and the overall tilt was different. When the shield reaches half of the building ($S_1 = -10$ m), the local slope value in each area is basically the same as the overall slope value because the building has been tilted as a whole, and the curve is attached to the axis x. In the second half of shield mining, the curve of Angle variables is symmetrical.

By comparing the vertical axis of Figs. 6.41, 6.42, 6.43, 6.44, 6.45 and 6.46, it is found that the value of the maximum Angle variable shows a significant downward

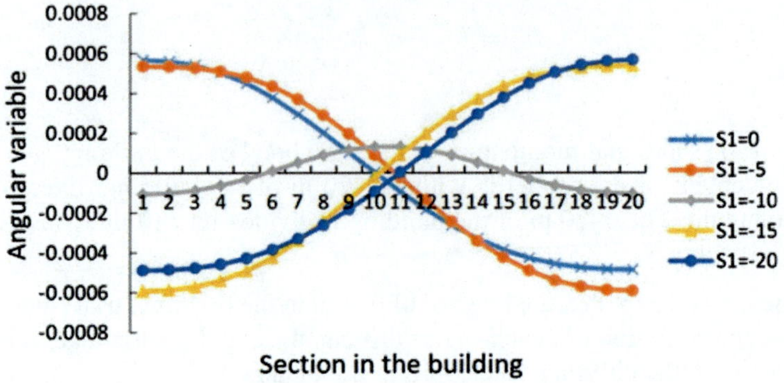

Fig. 6.41 Schematic diagram of various interval angle variables of the building under synchronous driving

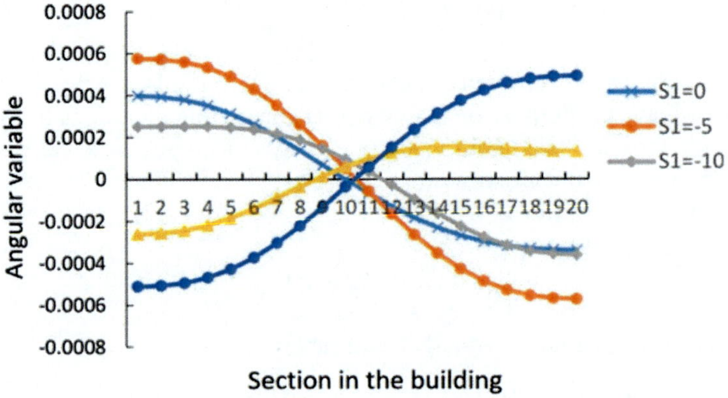

Fig. 6.42 Is a schematic diagram of angle variables of each interval of the building which are 1*D* apart before and after tunneling

trend, indicating the relationship between the distance between two shield machines and the local tilt of the building in the double-line shield. Therefore, in order to ensure the safety of the building, its tilt control requirements are higher, the distance between the two-line shield must take a larger value.

3) The maximum Angle variable under different working conditions is finally obtained

The maximum Angle variable during synchronous propulsion is $\beta_{\max 0} = 0.0565\%$; When the distance between front and rear driving is 1*D*, $\beta_{\max 1} = 0.0575\%$; when the driving distance between front and rear is 2*D*, $\beta_{\max 2} = 0.0364\%$; when the distance between front and rear tunneling is 3*D*, $\beta_{\max 3} = 0.0301\%$; when the distance between

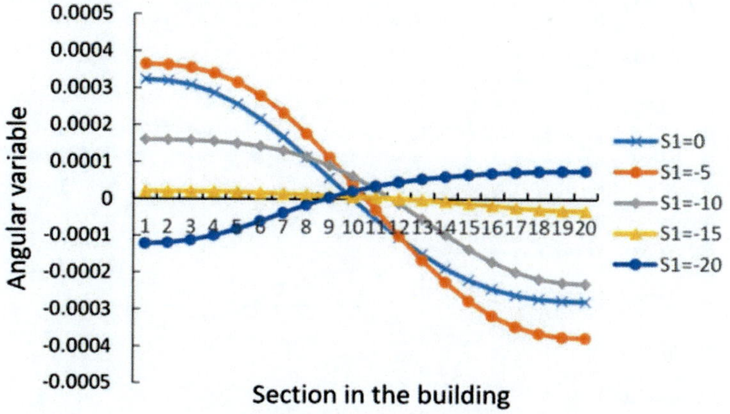

Fig. 6.43 Is a schematic diagram of angle variables of each interval of the building before and after 2*D* tunneling

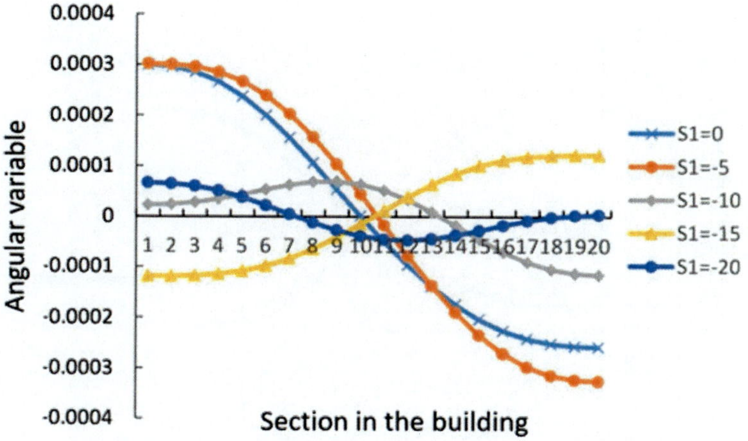

Fig. 6.44 Is a schematic diagram of angle variables of each interval of the building before and after 3*D* tunneling

front and rear driving is 4*D*, $\beta_{\max 4} = 0.0290\%$; when the front and rear drifts are 5*D* apart, $\beta_{\max 5} = 0.0287\%$.

(3) Analyze the building with safety control standards

The cracking of buildings due to settlement is related to many factors, including the mechanical properties of foundation soil, foundation type, structural materials, structural type and volume, distribution and size of the load on the structure, uniformity and velocity of settlement, etc. Due to the numerous influencing factors, the mechanism of building damage due to settlement becomes very complicated, so it is difficult to use the theoretical analysis method to obtain the allowable settlement of

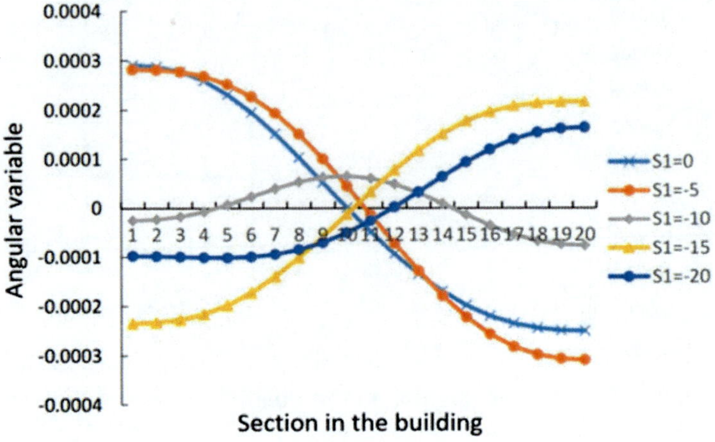

Fig. 6.45 Is a schematic diagram of angle variables of each interval of the building tunneling 4*D* apart before and after

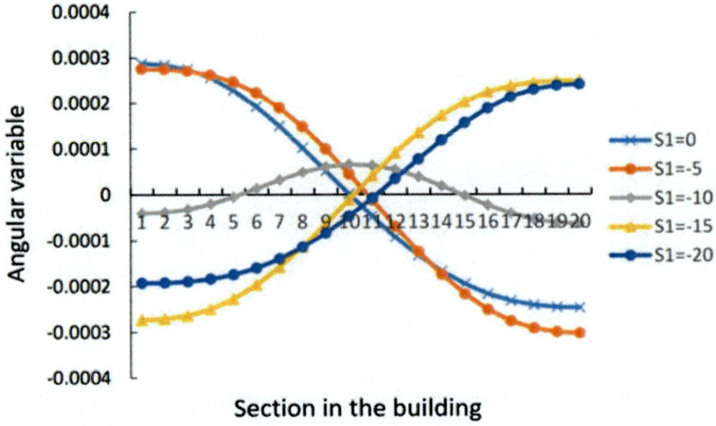

Fig. 6.46 Is a schematic diagram of angle variables of each interval of the building tunneling 5*D* apart before and after

the building. Therefore, the current standards on the permissible settlement of buildings are based on the observation of the settlement and damage of existing buildings on site.

The building mainly produces settlement due to its self-weight, and its horizontal displacement is very small and can be ignored. Therefore, the damage of the building under this circumstance is mainly related to the Angle variable and deflection ratio. Early scholars such as Kand and Peck [32], Polshin and Tokar [33]. Based on the observed data, some relationships between the building damage and Angle variables are given. Bjerrum [34] on the basis of previous studies and combined with relevant

Table 6.3 Relationship between angle variables and the degree of building damage

Angle variable β	Degree of building damage
1/750	Difficulty in operation of machinery sensitive to sedimentation
1/600	Hazard to frame structures with diagonal braces
1/500	Safety limits for buildings that do not allow cracks to occur
1/300	The partition wall began to crack
1/300	The operation of the crane was difficult
1/250	Rigid tall buildings begin to tilt noticeably
1/150	There are quite a few cracks in the spacer and brick walls
1/150	Safety limit of flexible brick wall (wall aspect ratio L/H > 4)
1/150	Buildings cause structural damage

observation data, the relationship between building damage and Angle variables is summarized as shown in Table 6.3.

Later scholars such as Burland and Wroth [35] and Grant et al. [36], Boscardin and Cording [37]. The allowable settlement of buildings has also been studied in succession, but the results obtained are basically not different from the values in Table 6.3. Table 6.3 is suitable for reinforced concrete frame and brick-concrete structures located in any soil layer, as well as buildings on independent foundations or raft foundations [38].

According to the maximum Angle variable calculated $\beta_{\max} = 0.0575\% < 1/750$, it can be seen that during the construction process of double-line shield mentioned in Sect. 6.2, the settlement problem occurred in this example did not have a great impact on the adjacent buildings. It can be seen that compared with the safety control standard that considers the maximum slope (corner), the standard of control Angle variable is more in line with the actual situation because it divides the building into smaller sections and considers the tilt, and combines the rigid body rotation momentum of the building itself, which is worth popularizing and using in the construction of backing structures.

6.6 Summary

In this book, with the help of the building foundation, shallow foundation and building synergy model above the tunnel with double parallel shield and double circular shield, the variation rules of building settlement and internal force in the tunneling area are analyzed, and the following conclusions are drawn:

(1) The influence of double-line parallel shield tunnel on the shallow foundation frame building indicates that the bending moment increases gradually with the approach of the excavation face. When the excavation face reaches the vicinity directly below the building, the bending moment reaches its maximum value and then decreases gradually, and the whole curve is approximately symmetric. The beam-column bending moment of the upper frame structure of the building is mainly determined by the tilt of the building. The greater the tilt rate is, the greater the constrained bending moment of the beam end and the bottom column on the foundation will be. Therefore, in the construction process, it is necessary to strengthen the monitoring of settlement difference and tilt rate at the end of the building, so as to better control the influence of double-line parallel shield tunneling on the bending moment of the adjacent shallow foundation frame building. Compared with the simultaneous tunneling, the two-line parallel shield tunneling successively has smaller tilt of the building, more moderate settlement change, smaller bending moment of the frame structure and foundation beam, and the distance of the excavation face increases with successive tunneling, but the maximum value is significantly smaller than the condition of simultaneous tunneling.

(2) Double-shield tunneling has a great impact on the adjacent shallow foundation frame buildings, and the buildings show an overall inclination. With the tunneling of the shield tunneling machine, the subsidence variation trend from the left end to the right end is relatively consistent. When the excavated surface of the shield machine reaches the vicinity of the building beam, the bending moment and shear force of the beam reach the maximum value, and then the bending moment and shear force gradually decrease, and the whole curve is approximately symmetric. In addition, in the construction process, it is necessary to strengthen the monitoring of the settlement difference and tilt rate of the building head and tail, so as to better control the influence of shield tunneling on the internal force of the adjacent shallow foundation frame structure. There are many factors affecting the deformation of frame buildings in the double shield tunneling area, among which the tunnel buried depth and soil loss rate have a particularly obvious influence on the buildings, and the smaller the tunnel buried depth, the greater the soil loss rate, and the greater the additional deformation and internal force of the buildings directly above the tunnel. The research shows that if the tunnel burial depth cannot be changed, the shield construction technology must be controlled reasonably, such as strictly controlling the grouting behind the tail wall of the double round shield.

(3) The building deformation control standards studied in this chapter adopt factors such as settlement control, slope control and Angle variables. Compared with the safety control standard that considers the maximum slope (corner), the control standard of Angle variable is more detailed in theory and in line with the actual situation, because it divides the building into small areas and considers the slope, and combines the rigid body rotation momentum of the building

itself. But in actual engineering, buildings and the model of idealized Settings is not the same, and the uneven settlement of existing buildings within has occurred, resulting in shield tunnel construction in the process of dynamic, if too much cause uneven settlement within buildings construction disturbance intensifies, will be a threat to the building structure security, even into dangerous state.

References

1. Peck RB. Deep excavations and tunneling in soft ground. In: Proceeding of 7th international conference on soil mechanics and foundation engineering. Mexico City: State of the Art Report; 1969, pp. 225–290.
2. Hu B, Liu Y, Tang H, et al. Research on ground subsidence due to tunnel excavation in Huquanmingdu section of Wuhan subway. Chin J Rock Mech Eng. 2012; 31(5):908–913.
3. Wei G, Pang S. The definition of close range between parallel shield tunnels based on numerical simulation. Municipal Eng Technol 2014; 32(1):76–80.
4. Han C, He G, Wang G. Analysis of surface settlement induced by some factors in parallel dual-tunnel construction. Rock Soil Mech. 2011; 32(S2):484–487, 495.
5. Ling H, Qiu W, Sun B, et al. Study of adjacent construction of two tube shield tunnels by centrifugal model test. Rock Soil Mech. 2010; 31(9):2849–2853.
6. Wei G. Predictionof soil settlement caused by double-line parallelshield tunnel construction. Disas Adv. 2013;6(6):23–7.
7. Wei G, Pang S. Study of three-dimensional soil deformation caused by double-line parallel shield tunnel construction. Rock Soil Mech. 2014;35(9):2562–8.
8. Peng C, Ji Y, Luo H. Numerical simulation of effects of double-tube parallel shield tunneling on neighboring building. Chin J Rock Mech Eng. 2008;27(S2):3868–74.
9. Li T, Chen H, Liu Bo, et al. Research on the influence of twin shield tunnel construction on the adjacent high rise building. J Hunan Univ Sci Technol (Nat Sci Edn). 2013; 28(4):43–48.
10. Wei G, Wei X. Effect analysis of frame building crossed by double-line parallel shield tunnel. Chin J Undergr Space Eng. 2013;9(2):339–43.
11. Hong J. Study on DOT shield tunnel construction disturbance and its impact on surrounding structures. Hangzhou: Zhejiang University; 2013.
12. Zhu H, Zheng Y, Chen H. Characteristics of soil surface settlement for Double-O-Tube shield tunnel. J Archit Civil Eng. 2006; 23(2):62–67.
13. Wei G, Chen W, Wei X. Prediction of surface settlement induced by double-o-tube shield tunnel excavation. Rock Soil Mech. 2011;32(4):991–6.
14. Sun T, Zhang Q, Wei L, et al. Analysis of additional stresses of soil disturbance induced by propulsion of double-O-tube shield. Rock Soil Mech. 2008; 29(8):2246–s2251.
15. Xia J, Yuan Y, Dong Z. Mechanism study on subsoil-strap footing-framework interaction in mining subsidence area. Chin J Geotech Eng. 2007;29(4):537–41.
16. Sagaseta C. Analysis of undrained soil deformation due to ground loss. Geotechnique. 1987;37(3):301–20.
17. Wei G, Xu R. Prediction of longitudinal ground deformation due to tunnel construction with shield in soft soil. Chin J Geotech Eng. 2005;27(9):1077–81.
18. Suwansawat S, Einstein HH. Describing settlement troughs over twin tunnels using a superposition technique. J Geotech Geoenviron Eng. 2007;133(4):445–68.
19. Bai Y, Dai Z, Xu F, et al. Study of the influence of one shield passing another on ground deformation for construction of parallel tunnels. Chin Civil Eng J. 2011;44(2):128–35.
20. Sun T, Li H, LV H, et al. Characteristics of the surface displacement induced by Double-O-Tube shield tunneling. Chin Civil Eng J. 2009;42(6):108–14.

21. Wei G, Chen C. Numerical simulation of the effect of Double-O-Tube shield tunnel construction on adjacent masonry buildings. Indus Construct. 2012;42(1):117–22.
22. Sun Y, Guan F. Shield tunnel construction induced influence on the settlement of masonry buildings. China Railway Sci. 2012;33(4):38–44.
23. Ding Z, Wei X, Wei G, et al. Study and analysis of internal force induced by shield tunnel construction of adjacent structure. Rock Soil Mech. 2011;32(S1):749–54.
24. Wei G. Selection and distribution of ground loss ratio induced by shield tunnel construction. Chin J Geotech Eng. 2010;32(9):1354–61.
25. Sheng W, Ye G, Qiao B. Influence of backfill grouting on ground settlement in DOT shield. Chin J Undergr Space Eng. 2014; 10(1):201–205.
26. Zhang M. On controlling technique of ground settlement during the Double-O-Tube (DOT) shield excavation. China Municipal Eng. 2009; 142(5):54–55, 92.
27. Skempton AW, MacDonald DH. The allowable settlement of buildings. Proc Inst Civil Eng. 1956;5(3):727–68.
28. Burland JB. Assessment of risk of damage to buildings due to tunneling and excavations. In: Invited special lecture, the first international conference on earthquake geotechnical engineering, vo. 1, no. 1, 1995, p. 95.
29. Liu G. Foundation pit engineering manual. Beijing: China Construction Industry Press; 2009.
30. Ge S, Xie D, Ding W, et al. Undercrossing disturbance control criterion for shield tunnel with consideration of building existing deformation. J Tongji Univ (Nat Sci). 2011; 39(11): 1616–1621.
31. Qi T. Settlement characteristics of strata and buildings caused bymetro tunneling. Chin J Geotech Eng. 2012;34(7):23–32.
32. Kand T, Peck RB. Soil mechanics in engineering practice. New York: Wiley; 1948.
33. Polshin DE, Tokar RA. Maximum allowable non-uniform settlement of structures. In: Proceedings of the fourth international conference on soil mechanics and foundation engineering, vol. 1, no. 11, 1957, pp. 402–406.
34. Bjerrum L. Allowable settlements of structures. Proc Eur Conf Soil Mech Found Eng. 1963;2(1):135–7.
35. Burland JB, Wroth CP. Settlement of buildings and associated damage. Proc Conf Settlement Struct. 1974;1(1):611–54.
36. Grant R, Christian JT, Vanmarcke EH. Differential settlement of buildings. J Geotech Eng Divis. 1974;100(9):973–91.
37. Boscardin MD, Cording EJ. Building response to excavation-induced settlement. J Geotech Eng Divis. 1989; 115(1):1–21.
38. Ou Z. Analysis and design theory and practice of deep excavation engineering. Taibei: Science and Technology Books Co., Ltd.; 2004.

Chapter 7
Prediction of Lateral Surface Settlement Caused by Shield Tunneling of Adjacent Buildings

7.1 The Introduction

The soil displacement and structure deformation caused by shield tunneling has been a common concern in urban subway construction in China, among which the peck formula is still the most widely used one to predict the lateral deformation of soil mass. Based on the long-term observation of surface settlement trough after tunnel construction and the analysis of a large number of measured data, Peck [1] proposed the normal distribution rule of surface lateral settlement trough during tunnel construction period, but the formula does not consider the existence of buildings. Most scholars still use peck's calculation formula for comparative analysis in their studies on the interaction of tunnel-build-soil [2, 3]. Ding believes [4] that ignoring the weight of the building itself will lead to significant changes in the width of the settlement trough caused by shield tunneling. And in practical engineering, due to the effect of the structural stiffness of the building, the deformation curve of the building is completely different from the natural foundation love, as shown in Fig. 7.1.

It can be seen from Fig. 7.1 that the deformation curve of the building is significantly different from the peck calculation formula curve and does not have a normal distribution. Based on this, Chinese scholar Han Xuan [6] improved the peck formula, and the Gauss curve based on the width of the settlement trough was proposed to fit the measured settlement curve. However, he adopted a larger parameter value of the width of the settlement trough in the fitting process, which was much higher than the recommended value of the width of the settlement trough in London clay. This book believes that it has not found that the settlement curve caused by the tunnel construction of the adjacent buildings can be skewed distribution, so the value can not accurately reflect the law of ground settlement caused by the tunnel construction.

At the same time, when the tunnel passes under the building, the settlement curve is very different from the peck formula prediction curve. The settlement curve is not normally distributed funnel-shaped, but symmetrically distributed cork-shaped curve. For example, Potts, etc. [7] proposed the "relative stiffness method" was

Z. Ding et al., *Influence of Shield Tunneling on Adjacent Structures and Control Technology*, https://doi.org/10.1007/978-981-19-1134-7_7

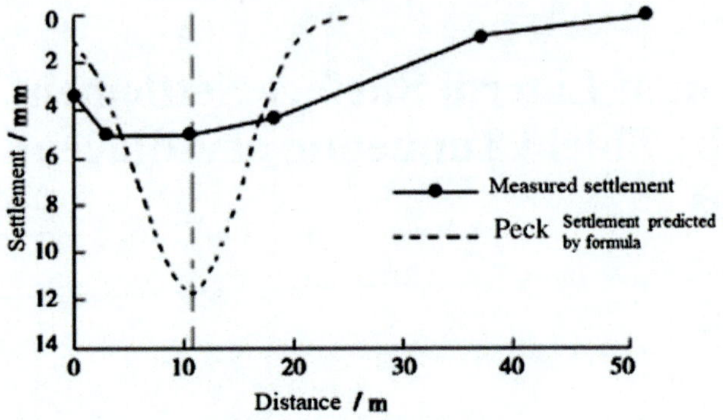

Fig. 7.1 Deformation prediction and measured settlement curve of London Mansion House [5]

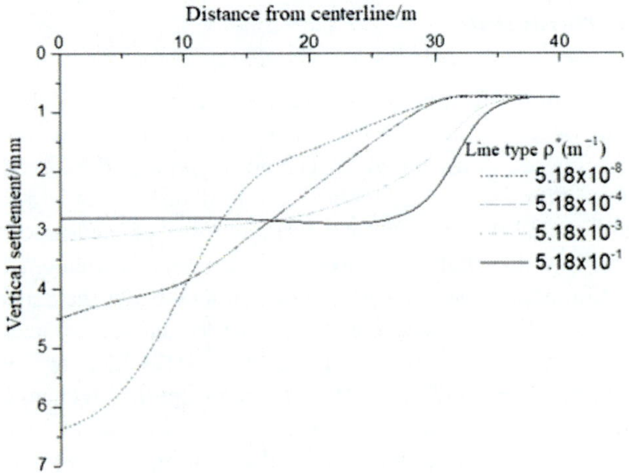

Fig. 7.2 Settlement curves of different structures under relative flexural stiffness

proposed by using the plane finite element method for numerical analysis. When the stiffness was zero, the settlement curve was normally distributed, while when the stiffness increased, the curve presented a "plug shaped curve" (see Fig. 7.2). Wang tao [8] established a three-dimensional finite element model, and the surface settlement figure obtained is shown in Fig. 7.3. When the tunnel passes under the building, its surface settlement does not have a normal distribution, but presents a "plug shaped curve" distribution as well. Not only the stiffness of the building restricts the displacement of soil, but also the distance between the tunnel and the building affects the distribution of settlement curve. Tunnel excavation has a certain influence range on soil displacement, as shown in Fig. 7.4. The soil is small when displacement beyond a certain range.

It can be seen that the existing studies on the distribution law of lateral surface settlement caused by subway tunnelling of adjacent buildings are relatively few, and

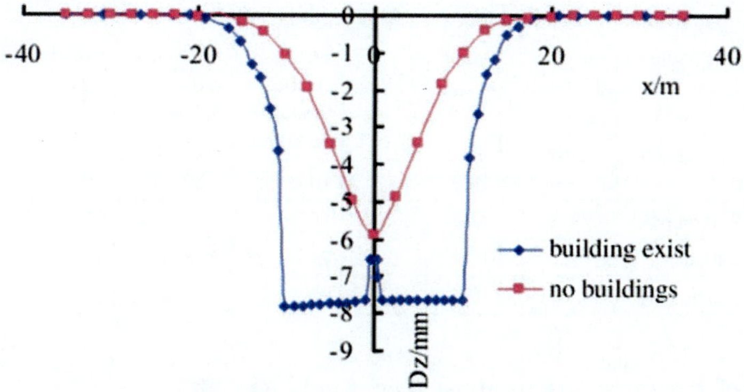

Fig. 7.3 Ground subsidence of buildings directly below the tunnel

Fig. 7.4 A schematic diagram of disturbance area of shield construction [9]

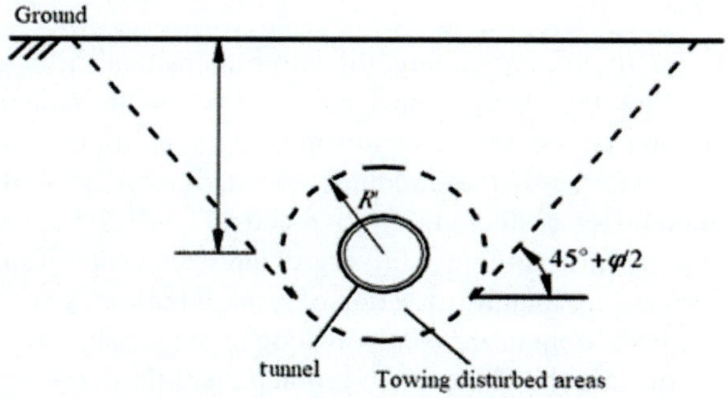

the existing empirical formula for predicting settlement "Peck formula" does not take into account the existence of buildings and their stiffness, nor the influence of buildings on settlement distribution curve. Based on my previous numerical results of the study, this chapter analyzes the adjacent building law of soil displacement caused by shield tunnel construction, puts forward the tunnel within the scope of the building directly, disturbance and outside disturbance range in construction of three kinds of working conditions, the surface subsidence, respectively is "plug shape distribution curve", "skewness distribution curve", and "the normal distribution curve" features, and presents a plug shape distribution curve calculation formula and the formula of skewness distribution curve and related parameters. And by establishing a permissible bending and deformation control of the tilt surface settlement building and permissible soil loss, took into account the different types of building structure type and foundation by using Delphi7.0 visual development tools, on the basis of research achievements in the numerical model using the Delphi software dynamic data binding technology, the system USES the Access as the database, visualization software is compiled "adjacent building subway shield tunnel construction system", to judge whether access shield tunneling horizontal sedimentation tank results in the adjacent building damage, to determine the damage of the buildings.

Table 7.1 Physical and mechanical indexes of each soil layer

The serial number	Soil name	Layer thickness	Volumetric weight/(kN/m ^{3})	The elastic modulus of/MPa	Poisson's ratio	Cohesion /kPa	Internal friction angle
1	Fill	4	18.5	10	0.30	10	12
2	Silty clay	8	18.5	15	0.33	12	20
3	Silt soil	8	17.1	7	0.40	11	8
4	Plastic clay	20	19.1	20	0.32	38	38

7.2 Preliminary Numerical Research Results

7.2.1 Model Establishment

In the preliminary study, the two-dimensional finite element method was used to simulate the shield tunnel construction of the adjacent buildings, and the beam-column system was used to simulate the buildings. Considering the different foundation forms of the buildings, the transverse ground settlement caused by shield tunneling was studied [10]. In practical construction, the interaction between tunnel and ground building is a very complicated three-dimensional problem. However, considering that the direction of ground buildings and tunnels is basically the same, and both of them are relatively long in the longitudinal direction, it can be simplified to consider the plane strain problem, and the section perpendicular to its strike can be taken as the calculation section [10], the two-dimensional model can also get a better solution [11]. See Table 7.1 for soil parameters of the model; see literature [10] for grid division, building and lining parameters; see Figs. 7.5 and 7.6 for building types and tunnel dimensions; H is tunnel overburden depth (m). L is the horizontal distance between the building axis and the tunnel axis (m); D is the outer diameter (m) of the shield.

7.2.2 Ground Settlement Analysis

Figures 7.7 and 7.8 numerical results show that [10], due to the presence of buildings, the land settlement curve no longer presents an axisymmetrical distribution, and the land settlement curve at the presence of buildings has obvious changed. When $L/D = 0$, the building is relatively safe and the surface settlement curve presents a plug-shaped distribution. When $L/D = 0.5$ to 2, the settlement difference between the head and the tail of the building is large, the building is relatively dangerous, and the surface settlement curve presents a skewed distribution. When $L/D = 3$ to 5, the impact of tunnel construction on buildings is small, and the surface settlement curve tends to be a normal distribution. When $L/D \geq 5$, the impact of tunnel construction on buildings can be ignored.

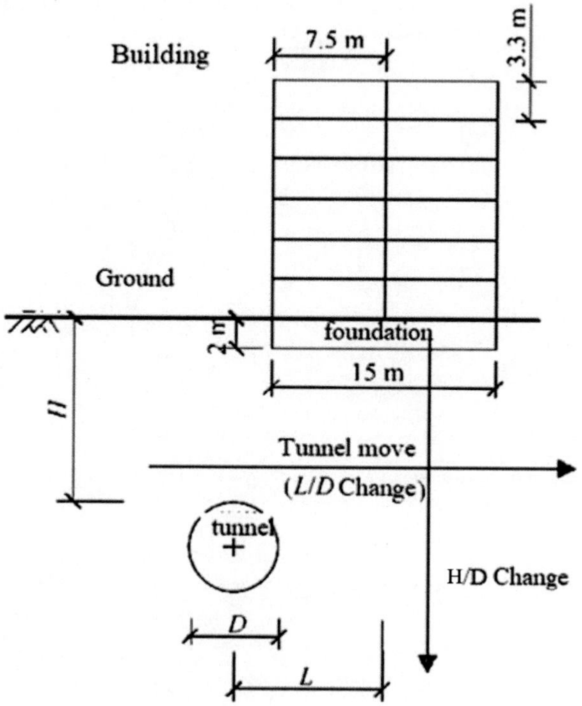

Fig. 7.5 Overall basic building and tunnel geometry

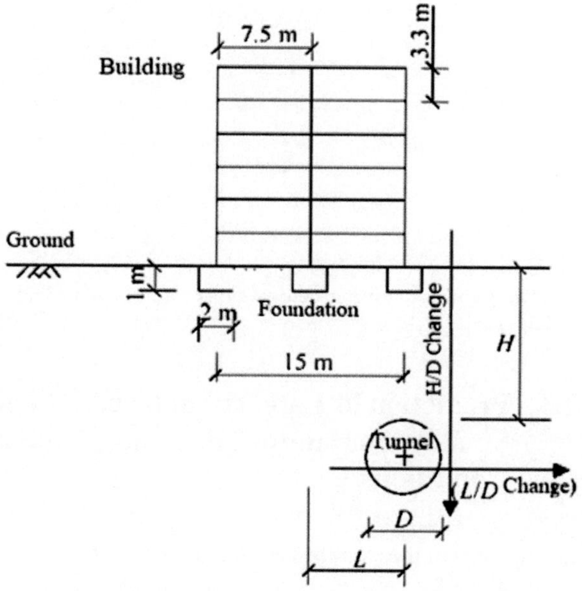

Fig. 7.6 Geometric relationship between independent foundation building and tunnel

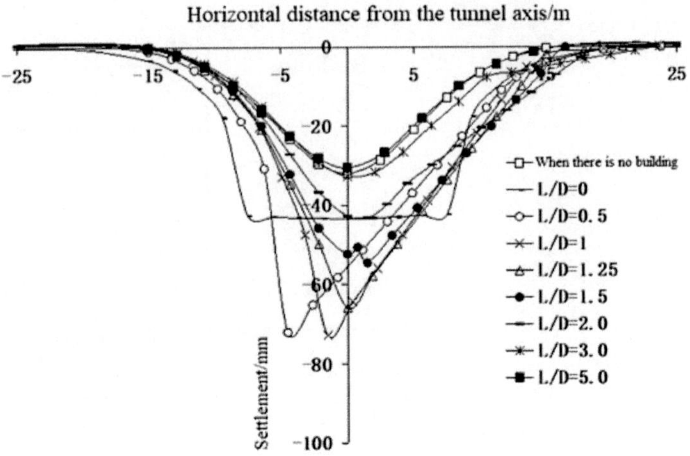

Fig. 7.7 Ground lateral deformation caused by tunneling of adjacent integral foundation buildings

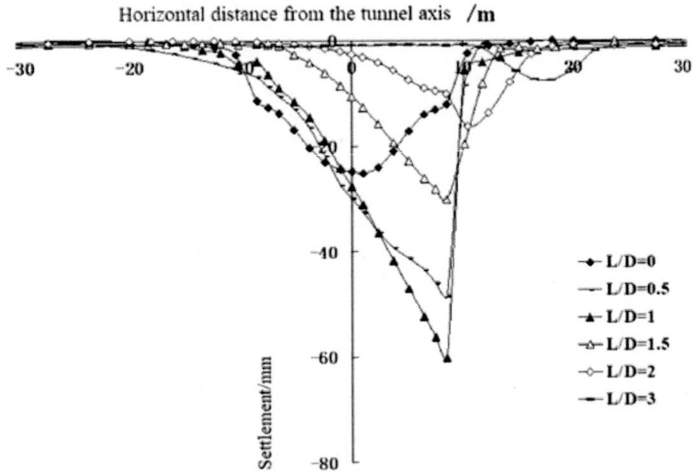

Fig. 7.8 Ground transverse deformation caused by tunneling of adjacent independent foundation buildings

7.3 Prediction of Lateral Surface Settlement of Adjacent Shallow Foundation Buildings Under Working Conditions

On the basis of the existing research in the previous section, this book puts forward the concept that the settlement curve can be in the shape of plug distribution and skewed distribution: (1) Tunnel excavation is under the building when the center of tunnel excavation crosses the axis of the building directly, it will cause the overall subsidence of the building, but the damage of tunnel construction to the building is relatively small. The settlement curve is a "plug curve" within the length of the building. (2) The tunnel is excavated near the building (the horizontal distance between the axis of the building and the axis of the tunnel is about 0.5–3 as the ratio of L to the outside diameter of the shield D). In this case, the ground settlement increases and the building cracks or even tilts. The settlement curve is "skewed curve". (3) A certain distance between tunnel excavation and buildings ($L/D \geq 3$) The ground settlement curve is relatively regular, and the settlement difference between the end and the end of the building foundation can almost be ignored. Therefore, the presence of buildings has little impact on tunnel excavation at this time, and the settlement curve is similar to the "normal distribution curve" of Peck formula.

7.3.1 Plug Distribution Curve

The stiffness of building structure will restrain the deformation of soil. For a building with uniform stiffness distribution, its constraint on soil deformation is also uniform and continuous, so the deformation curve of the building is also continuous, and the position of the maximum settlement point remains unchanged (or can be approximately regarded as unchanged in practical engineering) [12]. When the tunnel begins directly under the building, the foundation deformation becomes more uniform under the constraint action, that is, the settlement trough becomes more gentle and changes from a normally distributed funnel to a bottle stopper.

Celestino etc. [13] On the basis of the measured data, it is believed that the Peck formula has some errors in fitting, so he proposed the following prediction formula:

$$S = S_{\max} \frac{1}{1 + \left(\frac{|x|}{a}\right)^b} \tag{7.1}$$

where, $S_{\max}$: the maximum lateral ground settlement; a: Constant (length dimension), which affects the width of settling tank; b: A constant (dimensionless) greater than 1, which affects the settlement groove shape.

This book believes that its formula can well describe the "plug curve", which can be used to predict the ground displacement caused by the tunnel passing through the construction under the building. In this book, the values of parameters a and b in the equation are stipulated as follows:

When the distance between the settlement point to be solved and the center of the building's axis x is less than half of the foundation width B, $a/D = 0.8(z_0/D) + 0.5$, and when the distance between the settlement point to be solved and the center of the building's axis x is greater than half of the foundation width B, $a/D = 0.46(z_0/D) + 0.42$, the value of the parameter b is 2–3, which is related to the form of the foundation and the width of the building. Generally, the overall foundation is taken as a larger value.

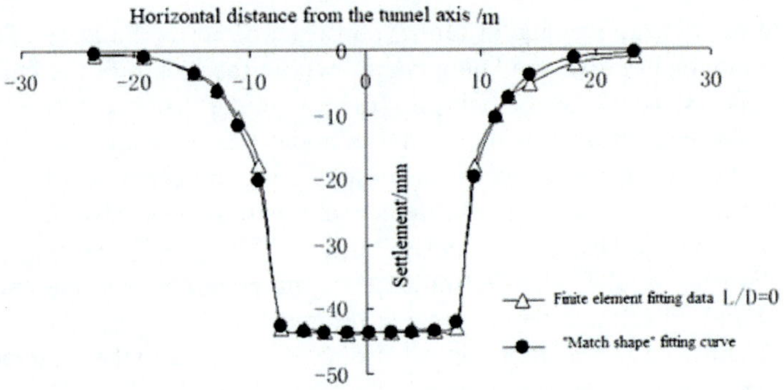

Fig. 7.9 Ground settlement Fitting Diagram caused by tunnel excavation (the tunnel is located directly under the building)

Two numerical analysis examples were used to verify the prediction formula. Example 1 was compared with the result when $L/D = 0$ in the numerical model of the whole foundation building in the previous section. As can be seen from Fig. 7.9, due to the existence of building stiffness, the finite element calculation results are significantly different from the normal distribution of peck formula. Although the distribution is symmetrical, the overall subsidence of the ground surface is shown due to the constraints of the building. As shown in Fig. 7.9, the shape and width of the settling groove are consistent. As the building foundation is an integral foundation, parameter b takes a larger value of 3.

In example 2, based on the finite element numerical simulation results from Maleki et al. [14], PLAXIS 3D, a three-dimensional finite element program, was used to analyze the interaction between the tunnel and the adjacent buildings, and the buildings were simplified to equivalent beams considering the different levels of the buildings. The dimensions of the tunnel and the building are shown in Fig. 7.10. The buried depth of the tunnel is 13.15 m, and the tunnel passes directly under the building. See the literature for the relevant physical parameters of the building, tunnel and soil [14].

It can be seen from Fig. 7.11 that the surface settlement caused by tunnel excavation directly under the building conforms to the "plug curve", which can well reflect the influence of the presence of the building on the displacement caused by tunnel construction directly below. In Fig. 7.11 fitting curve, parameter b is 2. Compared with Fig. 7.9, the width of settlement trough is consistent with the width of the building, but there are some differences in the settlement distribution shape, mainly because the literature [10] considers the overall foundation form. Generally, the value of a and b is related to the buried depth of the tunnel, the width of the building, the number of floors, the structural stiffness, and the soft and hard soil, etc. For the building with a large width, the value of b can be too small, and the value of b should be too large when the width is small. If the soil is soft, b can be taken to the larger value.

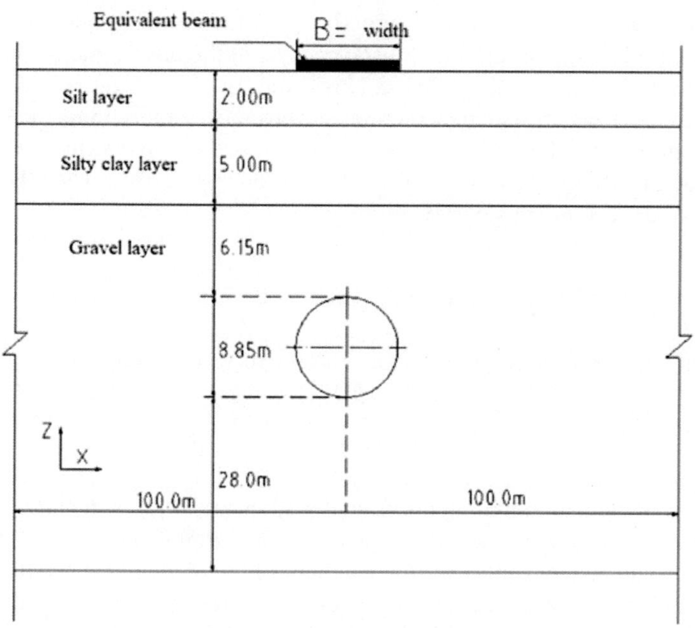

Fig. 7.10 Geometric relationship between building and tunnel

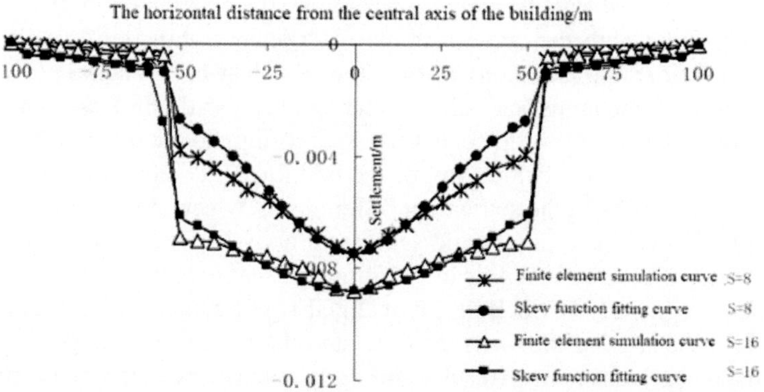

Fig. 7.11 Ground settlement fitting diagram caused by tunnel excavation under buildings with different floors

7.3.2 *Skew Distribution Curve*

As most subway tunnels do not pass directly under the buildings, the plug-shaped curve cannot fit the corresponding settlement curve completely. In this book, it is believed that within a certain range, buildings and tunnels are affected areas, and within the affected areas, the settlement distribution curve generally does not meet the normal distribution form.

When the tunnel is not tunneling directly under the building, but still within the affected area (the ratio of horizontal distance L from the building axis to the tunnel axis to the outside diameter D of the shield is about 0.5–3, the ground settlement will still increase, and the building may incline or even crack. At this point, the settlement curve is neither normal nor plug shaped curve, but a “skewed distribution curve”.

This book proposes the following skewness prediction curve: when the tunnel is to the left of the central axis of the building,

$$S(x) = S_{\max} \cdot e^{\frac{-\left[\ln \frac{x+\alpha \cdot B}{2L}\right]^2}{2w^2}} \tag{7.2}$$

When the tunnel is at the right of the central axis of the building,

$$S(x) = S_{\max} \cdot e^{\frac{-\left[\ln \frac{-x+\alpha \cdot B}{2L}\right]^2}{2w^2}} \tag{7.3}$$

where: $S_{\max}$ is the maximum lateral ground settlement; $S(x)$ is the land settlement distributed along the transverse direction; x is the horizontal distance from the settlement point to the tunnel center, m; B is the horizontal length of the building, m;L is the eccentricity, that is, the distance between the center of the tunnel and the center of the building, m; w is the empirical coefficient, which can be 0.60–0.70 according to the soil quality; α is the proportion coefficient, generally 0.8–1.0 which can be taken according to the building foundation form.

In order to verify the above prediction formula, the existing research results were respectively fitted with the curve formula of “skewness function”, and the fitting results also verified the rationality of the skewness distribution curve.

For example 1, the numerical calculation model and settlement calculation results in literature [10] were still adopted, w was 0.6 according to the soil quality and α was 1.0 according to the foundation form of the building. It can be seen from Fig. 7.12 that when $L/D = 0.5-1$, the settlement difference between the front and the rear of the building is large, and the ground settlement above the tunnel deviates from the normal distribution curve significantly, forming a large central subsidence area in the shape of a funnel. The building is inclined significantly. The settlement curve conforms to the skewed distribution curve, and the fitting result is good. However, based on the previous research results, this book believes that there is a certain range of skewed distribution curve. When $L/D \geq 3$, the land settlement curve still conforms to the normal distribution, the settlement difference between the end and the end of the building foundation can almost be ignored, and the presence of the building has little impact on tunnel excavation.

Example 2 uses datas from Han Xuan [12], when analyzing the settlement curve law caused by subway excavation, he collected many measured settlement data of neighboring buildings from the JLE Subway project in the United Kingdom, but he still adopted the Gaussian curve in the fitting, believing that the distance between the building and the tunnel has no great influence, which is obviously not consistent with the actual situation. Because it does not find that the settlement curve caused by

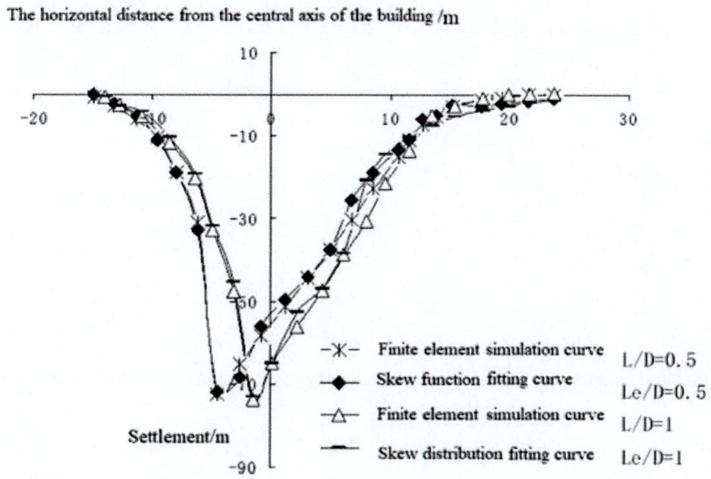

Fig. 7.12 A fitting diagram of ground settlement caused by tunnel excavation of the whole foundation building

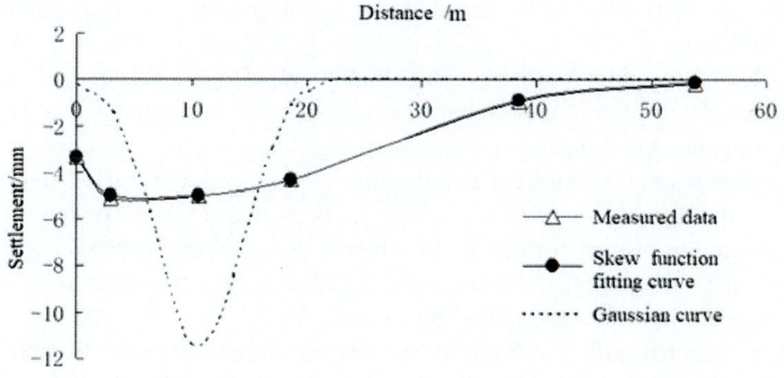

Fig. 7.13 A fitting figure for ground settlement caused by tunnel excavation at The Lord Mayor's Residence

tunnel construction of adjacent buildings can be skewed distribution, the fitting value cannot accurately reflect the surface settlement distribution form. In this book, two measured settlement curves adjacent to the Mansion House of The Mayor of London and Brick House are selected, and the curves are fitted by formula (7.2), as shown in Figs. 7.13 and 7.14, w is 0.6 according to the soil quality and α is 1.0 according to the foundation form of the building. The fitting results are in good agreement with the measured results, which further indicates the rationality of the fitting prediction of skewed distribution curve.

According to the measured settlement curves in Figs. 7.13 and 7.14, the surface settlement caused by tunnel construction of adjacent buildings is generally skewed distribution, and only when it is outside the influence range of tunnel excavation will

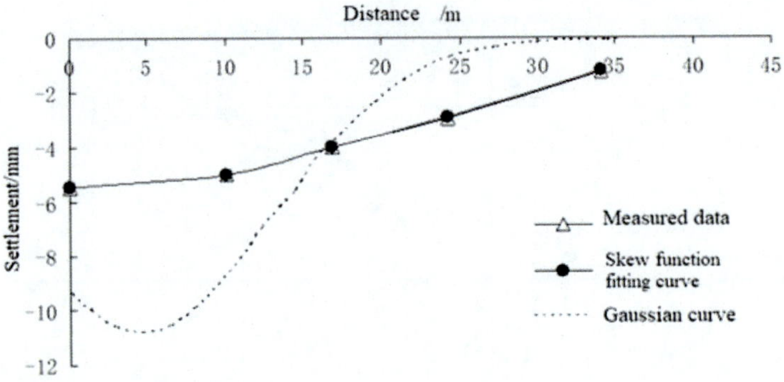

Fig. 7.14 Ground subsidence fitting caused by tunnel excavation in south wing of Brick Apartment Building

it obey the normal distribution of peck formula. The width and shape of the settlement trough in the skew distribution curve are determined by the distance, stiffness, soil quality and other factors of the adjacent buildings, which also indicates that the coupling action of soil, tunnel and buildings must be considered when studying the settlement distribution law caused by the construction of subway tunnel in the adjacent buildings.

Example 3 was calculated by using Shahin et al. [15] model test data. The centrifuge model test was conducted to study the surface settlement caused by tunnel excavation of adjacent buildings (as shown in Fig. 7.15), and the score was compared with two-dimensional numerical simulation. Where D is the buried depth of the tunnel; B is the driving range of the shield, L_P is the buried depth of building foundation. See literature [16] for details of relevant parameters. Figure 7.16 shows the settlement curve of the data obtained from model test, the curve obtained from finite element analysis and the curve fitted by formula (7.2).

It can be seen from Fig. 7.16 that the predicted curve of skewed distribution is in good agreement with the results of finite element numerical simulation and model experiment. w is set at 0.6 according to the soil quality and 0.8 according to the foundation form of the building.α However, compared with the prediction results in Figs. 7.13 and 7.14, the dispersion is slightly larger, which may be due to the depth of the building foundation. The settlement curve has a skewed distribution, but its slippage is slightly worse.

It can be seen that the soil movement caused by shield tunneling of adjacent buildings runs through the construction of the tunnel. In addition to the influence of soil quality, it is also affected by various factors such as the buried depth of the tunnel, change of groundwater level, grouting effect, distance of the building, number of floors, stiffness, foundation form and so on. In the past, it is not reasonable to study the surface settlement caused by tunnel excavation of adjacent buildings based on Gaussian curve or Peck formula. Therefore, the interaction among soil, tunnel and buildings should be considered comprehensively in the study of shield tunnel construction under the working condition of adjacent buildings. In general, when

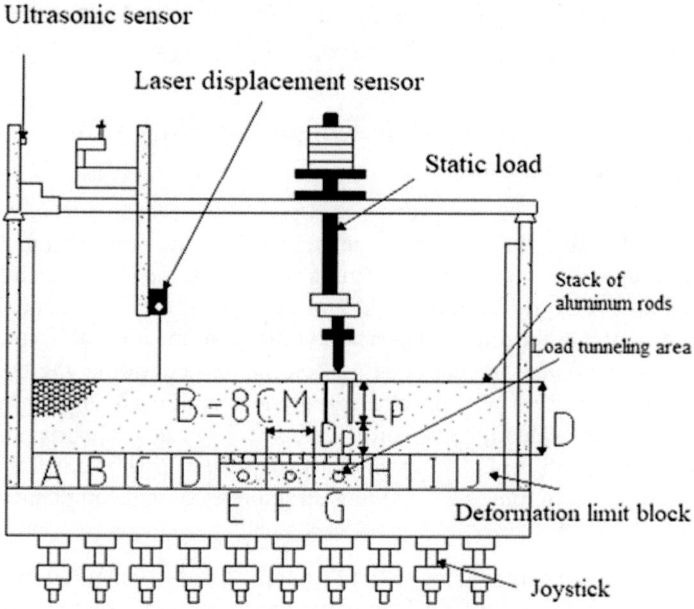

Fig. 7.15 Schematic diagram of tunnel excavation of adjacent buildings simulated by centrifuge experiment

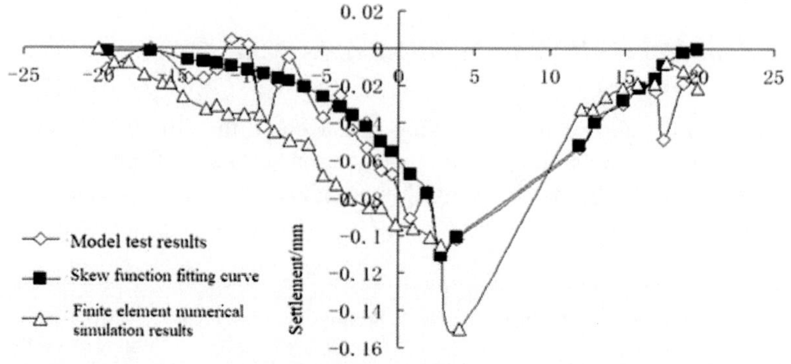

Fig. 7.16 Ground settlement fitting diagram for centrifuge experiment simulation

tunnel excavation is carried out under the building, within the disturbance range and outside the disturbance range, "plug curve", "skewness curve" and "normal distribution curve" can be considered for surface settlement respectively.

7.4 Development of Visual Software for Safety Assessment of Shield Tunneling of Adjacent Buildings

7.4.1 Evaluation Criteria of Sedimentation Tank Width Parameter Method

Based on the allowable surface settlement standard for the deformation control of the building foundation, the previous research results assumed that the building was a homogeneous elastic foundation beam and introduced the settlement slot width parameter K to obtain the allowable surface settlement, the allowable soil loss rate and the allowable tilt rate for the bending deformation control of the building, as shown in Tables 7.2 and 7.3 [17]. Based on the grade of building damage in the Jubilee MTR extension line evaluation standard (see Table 7.4), the evaluation standard of settlement trough width parameter method is established, as shown in Table 7.5. At the same time, the differences of building structure type and foundation type are considered.

Where: i is the width coefficient of settling trough; R is the tunnel radius.

And on the basis of the numerical simulation in the previous section, considering the difference of concrete structure and foundation, the building evaluation reduction coefficient is proposed, as shown in Table 7.6.

7.4.2 Realization of Visual Influence System

On the basis of the above analysis, taking into account the different structures and foundation types of the buildings, the visualization software "Tunnel Construction influence System of adjacent buildings" is compiled, which can further judge the

Table 7.2 Surface allowable settlement and allowable soil loss rate [18]

Damage level	Description of severity	Ultimate tensile strain ε_n (%)	Permissible settlement $[S_{\max}]_n$ (10^{-2} M)	Allowable soil loss rate $[\eta]_n$(%)
0	Almost negligible	0–0.05	0–0.320 i	0–0.255 $(i/R)^2$
1	Very mild	0.05–0.075	0.320 i–0.394 i	0.255–0.314 $(i/R)^2(i/R)^2$
2	Slight	0.075–0.15	0.394 i–0.568 i	0.314–0.453 $(i/R)^2(i/R)^2$
3	moderate	0.15–0.3	0.568 i–0.831 i	0.453–0.663 $(i/R)^2(i/R)^2$
4, 5	Serious to very serious	>0.3	>0.831 i	>0.663 $(i/R)^2$

Table 7.3 Permissible tilt rate of buildings [18]

Type of foundation soil	Allowable slant rate $[f]$ (%)	Permissible settlement $[S_{\max}]_n$ (10^{-2} m)	Allowable soil loss rate $[\eta]_n(\%)$
Medium and low compression soil	0.2	0.329 i	0.263 $(i/R)^2$
High compression soil	0.3	0.494 i	0.394 $(i/R)^2$

Table 7.4 The corresponding relationship between the damage level of the building and the ultimate tensile strain [18]

Damage level	Description of severity	Ultimate tensile strain (%)
0	Almost negligible	0–0.05
1	Very mild	0.05–0.075
2	Slight	0.075–0.15
3	Moderate	0.15–0.3
4, 5	Serious to very serious	>0.3

Table 7.5 Classification of visible damage to buildings [18]

Damage type	Damage	A description of typical breakage
0	Almost negligible	Fracture less than 0.1 mm
1	Very mild	The cracks are fine and can be disposed of by decoration. The damage usually occurs in the interior wall, and the typical crack width is less than 1 mm
2	Slight	Cracks are easy to fill and may need redecorating. Cracks are visible from the outside. Doors and Windows may tighten slightly; Typical cracks can be up to 5 mm wide
3	Moderate	Cracks need to be repaired, doors and Windows are difficult to open, water pipes or gas pipes may break, the waterproof layer is weakened, typical cracks up to 5–15 mm
4	Serious	In particular, the upper walls of doors and Windows may need to be chiseled off. The doors and window frames are twisted, the floor tilt can be perceived, the wall tilt or protruding can be perceived, the pipes are broken, and the typical crack width can reach 15–25 mm
5	Very serious	This project may require partial or complete reconstruction of the original house. The beam loses its bearing capacity, the wall is seriously inclined, the Windows are twisted, broken and the structure is unstable. The typical crack width is greater than 25 mm

Table 7.6 Building reduction coefficient [18]

Structure type	The base type	Reduction factor
Masonry structure	Independent foundation	1/3
	Continuous basis	1/2
	Raft foundation	1/2
	Pile foundation	3/4
Reinforced concrete structure	Independent foundation	2/3
	Continuous basis	1
	Raft foundation	1
	Pile foundation	3/2

damage degree of shield construction to adjacent buildings. Adjacent building tunnel construction affect system research and development of the overall design concept is to establish a more convenient control basis, first calculate the structure of the allowable tensile strain and tilt rate, then calculate the ground deformation caused by shield tunneling at baseline [$S_{\max}$], and then to the ground surface deformation as a basic value evaluation to determine whether deformation of adjacent buildings as well as the further construction of the damage.

(1) Delphi technology

The development tool of the system uses Delphi [19], which is a visual development tool developed by Borland Software Company, can be used in Windows and other system environment, it has a visual corresponding integrated development use environment (IDE). At the same time, it provides more than one hundred can use research and development of components, the use of these related artifacts developers can be a very quick constructs the ideal system, and contains more reusable components, and allow developers to control based on the development effect of Windows environment, and it also has very strong research and development into and read data function. Delphi technology can not only be used in the development of related system software, but also very suitable for the application of disciplines and software development.

(2) Dynamic data binding technology

Dynamic data binding technology refers to changing the data source into an observer, and then registering the control as a software observer into a private object such as the relevant data source. When the relevant state or data in the data source changes, it will actively inform all the bound controls, namely observers, and then the relevant controls will automatically extract the software data to complete the update. Once the related data source is built, the control can also carry out the latest update in the data source. Once a user USES such controls to make changes to the relevant state or data, such changes are naturally notified to all other relevant data controls in real time through the relevant data source. This is in fact the Push Mode for transitioning from data sources to controls.

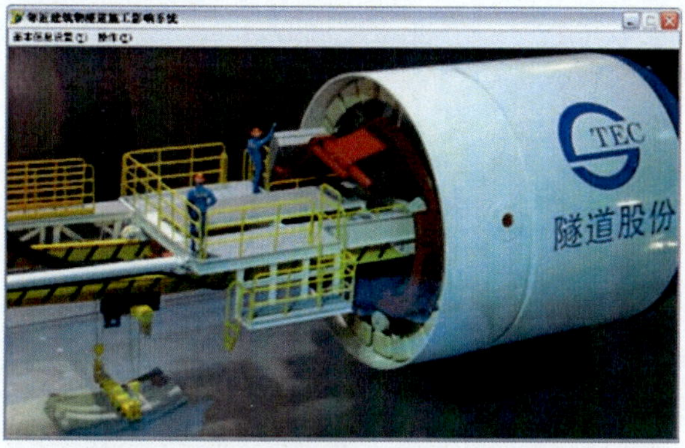

Fig. 7.17 Software interface

(3) Computer simulation interface

By using Delphi7.0 visual development tool, we can directly work in the visual development environment with good user interface. Under the control of the operating system, we can establish the calculation model based on the influencing factors of tunnel construction in adjacent buildings. On the basis of the model, the dynamic data binding technology of Delphi is used to easily obtain the basic data and carry out the calculation. After that, the calculation results are displayed intuitively. See the appendix of this paper for detailed programming procedures.

The specific interface is shown in Fig. 7.17:

The system USES Access as the database, which is used to configure the parameters such as the width of the settlement tank, the classification of damage degree, the allowable settlement parameters and the reduction coefficient, etc., to facilitate the selection according to various environments during calculation. The system uses a large number of dynamic data binding controls, which is conducive to the rapid implementation of data presentation and update.

For the calculation of damage classification, the system calculates the required result form according to the region and the basic formation characteristics of the region, building reduction coefficient, distance reduction coefficient combined with the input S value and H value according to different comparison methods.

Its main interface is shown in Fig. 7.18:

In the interface of tunnel construction influence system of adjacent buildings, the first choice is to determine the K value (the width parameter of settlement trough), by pressing the button "select area" and "select basic formation characteristics". After the K value is determined, the width coefficient of settlement i can be determined by the formula $i = K_0 h$, h is the buried depth of the tunnel. Secondly, considering the

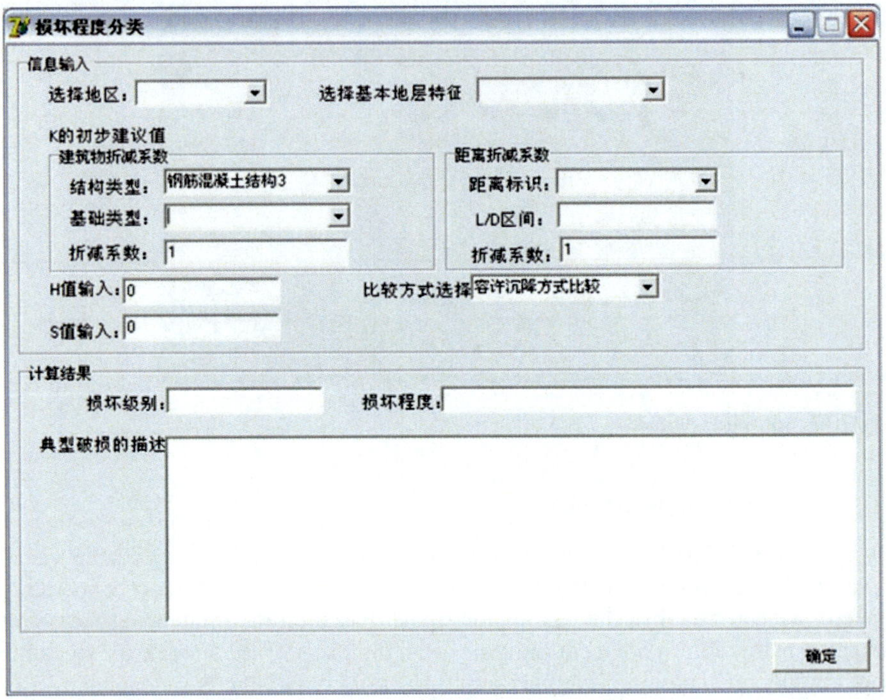

Fig. 7.18 Input interface

difference of building structure, foundation type and distance, the reduction coefficient is obtained. Then choose a way: "the permissible settlement way", "ultimate tensile strain mode comparison" and "allow soil loss compare", then input respectively "tunnel buried depth h (m)" and "settlement calculation values S (cm)" ("tensile strain calculated value "ε or" value of soil loss rate calculation η") and "tunnel radius R (m), end up with buildings visible damage, as shown in Fig. 7.19.

(4) Database design model

The database tables related to the system are designed as follows:

(1) Parameters of settling tank width

The serial number	The field name	The field type	The default value
1	cjckdxs_dq	The text	
2	cjckdxs_dctz	The text	
3	cjckdxs_k_value	Digital	

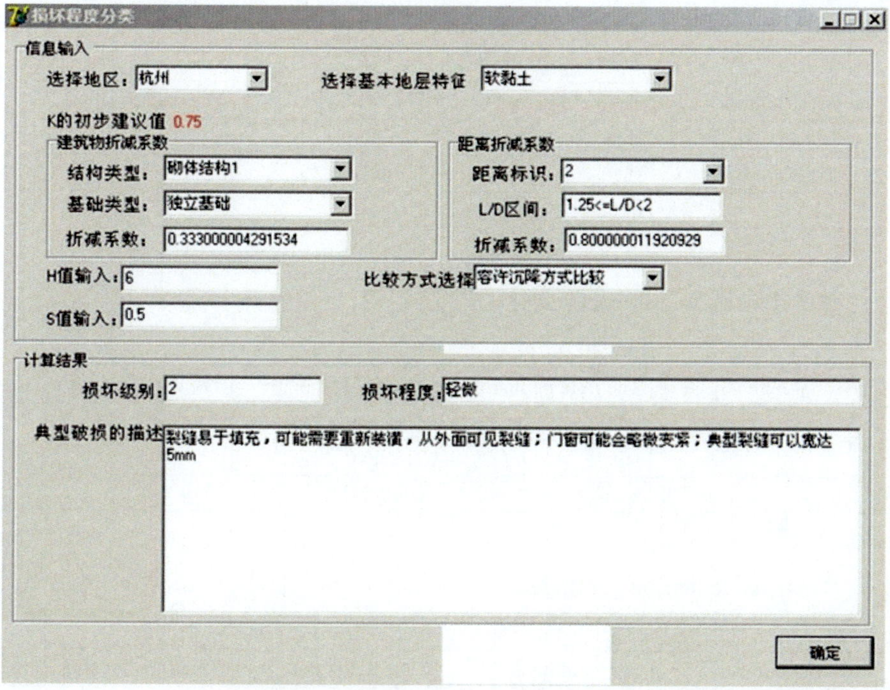

Fig. 7.19 Interface of calculation results

(2) Allowable settlement parameters

The serial number	The field name	The field type	The default value
1	shjb	Digital	
2	yzxms	The text	
3	jxlyb_min	Digital	
4	jxlyb_max	Digital	
5	rxcj_min	Digital	
6	rxcj_max	Digital	
7	rxrxttssl_min	Digital	
8	rxrxttssl_max	Digital	

(3) Damage degree classification table

The serial number	The field name	The field type	The default value
1	shjb	Digital	
2	shcd	The text	
3	bj	The text	

(4) Building reduction coefficient

The serial number	The field name	The field type	The default value
1	jglx	The text	
2	jclx	The text	
3	zjxs	Digital	

(5) Distance reduction coefficient

The serial number	The field name	The field type	The default value
1	jlbs	Digital	
2	ld_min	Digital	
3	ld_min_include	Yes/no	
4	ld_max	Digital	
5	ld_max_include	Yes/no	
6	zjxs	Digital	

7.5 Summary

Tunnel tunneling will inevitably cause a certain degree of ground disturbance to the circumfluence soil, and the influence of the surface lateral settlement trough on the adjacent buildings will be related to the size of the surface deformation, the relative position between the tunnel and the buildings, as well as the building foundation, structural type, and the formation characteristics:

(1) To study the shield construction in the building area, the correlation between soil, tunnel and the building should be considered comprehensively. The size of soil settlement trough caused by shield tunneling should consider the existence of adjacent buildings. Ignoring the influence of buildings will make the calculation of surface deformation caused by excavation small, thus increasing the

risk of construction. The influence of buildings on shield tunnel construction has a certain range, beyond which the influence of the existence of buildings on tunnel construction can be ignored. The horizontal distance between the tunnel and the axis of the building the buried depth of the tunnel and the different foundation forms of the building are all important factors that cause the surface deformation.

(2) The analysis and verification of calculation examples show that it is reasonable for the surface settlement to show "plug curve", "skewness curve" and "normal curve" respectively within a certain range. Tunnel excavation is under the building, and the center of tunnel excavation just crosses the axis of the building, which will cause the overall subsidence of the building. The settlement curve is "plug curve" within the length of the building. The tunnel is excavated near the building (the horizontal distance between the axis of the building and the axis of the tunnel is about 0.5–3 times the ratio of L to the outside diameter of the shield D), and the settlement curve is "skewed curve".The ground settlement curve of the tunnel excavation at a certain distance ($L/D \geq 3$) from the building is relatively regular, so the presence of the building has little influence on the tunnel excavation at this time. The settlement curve is similar to the "normal distribution curve" of Peck formula.

(3) According to the tunnel construction in our country in the process of building damage evaluation standard of blank, by the structure of the allowable tensile strain and tilt rate of the surface deformation at baseline, and adjacent to the ground surface deformation as a basic value evaluation to determine whether different distance building construction damage, can be used the subway tunnel, the tunnel design and construction of crossing the street. Using Delphi7.0 visual development tool, under the control of the operating system, the model was established to calculate the factors affecting tunnel construction of adjacent buildings, and Delphi dynamic data binding technology was used to obtain the basic data for calculation, and the results could be visually displayed. Using Access as a database, it is used to configure parameters such as width of settlement slot, classification of damage degree, allowable settlement parameter and reduction coefficient, etc., and a large number of dynamic data binding controls are used, which is helpful to realize the visualization of tunnel construction prediction of adjacent buildings quickly.

References

1. Peck RB. Deep excavations and tunneling in soft ground. Proceeding of 7th International Conference on Soil Mechanics and Foundation Engineering. Mexico City: State of the Art Report, 1969;225–290.
2. Mroueh H, Shahrour I. A full 3-D finite element analysis of tunneling-adjacent structures interaction. Comput Geotech. 2003;30:245–253.
3. Jenck O, Dias D. 3D-finite difference analysis of the interaction between concrete building and shallow tunneling. Geotechnique. 2004;54(8):519–528.

4. Ding Z, Wei X, Zhang T et al. Analysis and discussion on surface settlement induced by shield tunnel construction of adjacent sructure. Disaster Adv. 2012;5(4):1656–1660.
5. Frischmann WW, Hellings JE, Gittoes G. Protection of the mansion house against damage caused by ground movements due to the docklands light railway extension. Geotech Eng. 1994;107(2):65–76.
6. Han Y, Li N, Standing JR. An adaptability study of Gaussian equation applied to predicting ground settlements induced by tunneling in China. Rock Soil Mech 2007;28(1):23–28.
7. Potts DM. Addenbrooke T I. A structure's influence on tunnelling-induced ground movements. Geotech Eng. 110(2):109–125.
8. Wang T. Study on the influence of shield tunneling on surrounding environment. Zhejiang University Hangzhou; 2007.
9. Wei G. Research on theoretical calculation of long-term ground settlement caused by shield tunneling. Chin J Rock Mech Eng. 2008;27(S1):2960–2966.
10. Ding Z, Wei X, Wei G et al. Numerical analysis of surface settlement induced by shield tunnel construction of adjacent structure. Rock Soil Mech 2009;30(S2):550–554.
11. Xu J, Gu Y, Kang J. Study on interaction of tunnel and upper structure. Rock Soil Mech. 2005;26(6):889–892.
12. Han X. The analysis and prediction of tunnelling-induced building deformations. Xi'an: Xi'an University of Technology; 2006.
13. Qing W, Liao H, Qian C. The effect construction on the settlements of adjacent building and earth surfaces of underground tunnel. Chin J Underground Space Eng. 2005;1(6):960–963, 978.
14. Celestino TB, Gomes RAMP, Bortolucci AA. Errors in ground distortions due to settlement trough adjustment. Tunn Undergr Space Technol. 2000;15(1):97–100.
15. Maleki M, Sereshteh H, Mousivand M et al. An equivalent beam model for the analysis of tunnel-building interaction. Tunnelling Underground Space Technol. 2011;26(2):524–533.
16. Cording EJ, Hansmire WH, Macpherson HH. Displacement around tunnels in Soils University of Illinois Report Prepared for Department of Transportation, Urbana; 1976.
17. Shahin HM, Sung E, Nakai T et al. 2D model tests and numerical simulation in shallow tunneling considering existing building load. Underground Construction Ground Movement. 2006;15:67–82.
18. Ding Z. Research of interaction between shield tunnel and adjacent structure. Zhejiang University Hangzhou; 2007.
19. Chen S. Delphi deep exploration. Wuhan: Huazhong University of science and Technology. Press, 2004.

Chapter 8
Construction Control Technology of Shield Construction for Adjacent Structures

8.1 Introduction

Shield construction of adjacent structures will inevitably lead to a certain degree of stratum displacement, which will result in subsidence deformation of the surface and structures and endanger the safety of structures. Therefore, people have been concerned about how to reduce the disturbance of shield construction to the surrounding soil, minimize the influence of shield construction on adjacent structures, and the corresponding control and protection measures. The technical measures to control the subsidence deformation of ground and adjacent structures caused by subway shield construction can be mainly divided into three aspects: stratum and adjacent structures treatment measures, subway shield construction parameter optimization and reconstruction, and construction monitoring and measurement analysis.

(1) Treatment measures for strata and adjacent buildings

The influence of soil disturbance can be alleviated by improving the rigid response of the stratum, or the structure of blocking deformation can be set in the soil layer near the protected building, so as to reduce the settlement deformation of the stratum and adjacent buildings caused by shield construction. It mainly includes measures such as separation pile, partition wall, grouting, foundation support and building structural reinforcement.

(2) Optimization of metro shield construction parameters

The main control parameters for the tunneling of subway shield tunneling machine are earth bunker pressure, thrust and distribution of jack, advance speed, shield slope, deviation correction direction and amount, grouting method, quantity and grouting pressure, etc. Shield is the optimum combination of parameters in the construction process in order to control of shield tunneling axis deviation does not exceed allowed range and minimize the influence of stratum deformation, meanwhile must match with the corresponding monitoring method, all kinds of data will be measured

Z. Ding et al., *Influence of Shield Tunneling on Adjacent Structures and Control Technology*, https://doi.org/10.1007/978-981-19-1134-7_8

and the monitoring of surface subsidence value, to guide next step excavation, informatization construction.

(3) Monitoring measurement back analysis

Back analysis of monitoring and measurement plays an important role in practical engineering. By monitoring and measuring, according to the measured data analysis, feedback, correction of various parameters, correct construction guidance: 1) To understand the deformation of the stratum and the building around the shield construction, to provide basis for daily construction management; 2) Determine whether the construction technology and construction parameters meet the expected requirements, so as to determine and optimize the construction parameters and sequence, achieve dynamic information-based construction management, and provide basis for the modification of engineering design scheme; 3) Ensure the normal use of buildings, underground pipelines and other structures within the scope affected by the construction, so as to provide basis for the reasonable determination of protection measures.

In this chapter, three construction cases of adjacent buildings with different shield structures are listed, including the pile project of fengqi Bridge under the subway section of Jianguo North Road Station–Middle Hebei Road Station in Hangzhou, the project of Tunnel under the West part of Fenghuang Garden under the subway section of Jiaoqiao Station–Changjiang Road Station in Nanchang, and the project of tunnel under the important cultural relics under the section of Tunnel under the Drum Tower Station in Ningbo. By describing the shield construction control technology of adjacent buildings adopted in different projects, the concrete implementation methods of reconstruction of shield machine, self-reinforcement of building and shield tunnel, grouting of stratum, construction monitoring and measurement are given. By analyzing the settlement and deformation of the ground and buildings, the rationality and effectiveness of the engineering control measures can be evaluated, which can provide useful reference for similar projects.

8.2 Case 1: The Subway from Jianguo North Road Station to Middle Hebei Road Station in Hangzhou Passes Through Fengqi Bridge Pile

8.2.1 Project Overview

(1) Project overview

The interval from Jianguo North Road Station to Middle Hebei Road Station is the whole underground shield interval, with the upper line length 552.008 m and the lower line length 550.933 m. Three sets of plane curves are set in the whole interval, with the radius of the curves being 700 m, 1000 m and 2500 m respectively. One-way slope is adopted for the interval route less than 600 m, and the maximum longitudinal slope is 6.902‰. Vertical curve is set at the point of slope change, and the buried depth of the tunnel is between 10 and 11.7 m. The shield starts from Jianguo North Road station, arrives at Zhongbei Road Station and then transfers to Jianguo North Road Station for a second departure, with an east–west direction. It will cross the Fengqi Bridge, cover the East River with soil of 6.1 m, and be 4.6 m clear from the nearest building (Wansheng Garden).

(2) The engineering environment of the Fengqi Bridge section

The reformed German Herecke earth-pressure balance shield machine is used to grind piles and cross the river in the interval from Jianguo North Road station to Middle Hebei Road Station. Shield across the east river need grinding four piles, 2 lines in which are bored piles for Φ1000 mm, pile foundation concrete grade is for C25, and main reinforcement is for Φ22; another 2 lines is for Φ300mm (plum flower arrangement), reinforced concrete tree pile foundation concrete grade and the main reinforcement is unknown, implementation plan according to the main reinforcement of reinforced Φ16, according to the concrete C25 consideration. Pile up line grinding 4 root for Φ1000 mm; downlink grinding 6 root pile for Φ1000 mm. The ground piles are basically perpendicular to the tunnel line (Figs. 8.1 and 8.2).

(3) Geological conditions of crossing fengqi Bridge section

The section of shield tunneling through Fengqi Bridge is mainly composed of strata ③$_6$ Silty silt with sandy silt, ③$_7$ Sandy silty and ⑥$_1$ Silty clay. At the same time, the tunnel was covered with shallow soil, about 6.1 m. The overburden is mainly composed of ③$_6$ Silty silt with sandy silt, ③$_3$ Sandy silt with silt, ③$_2$ Sandy silty soil and a small amount of silt fill (Fig. 8.3, Tables 8.1, 8.2 and 8.3).

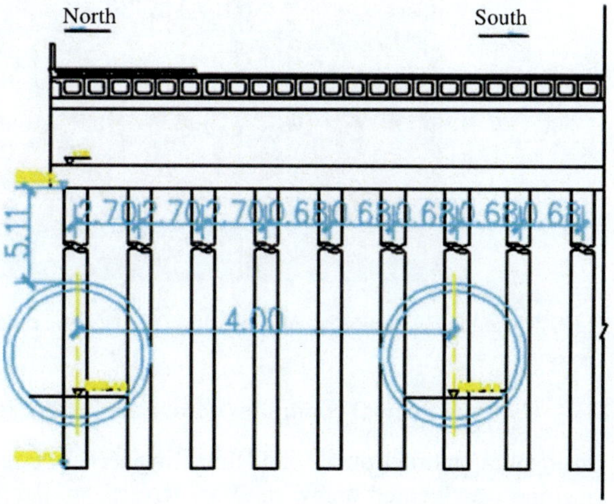

Fig. 8.1 Section diagram of the positional relationship between tunnel and Fengqi Bridge

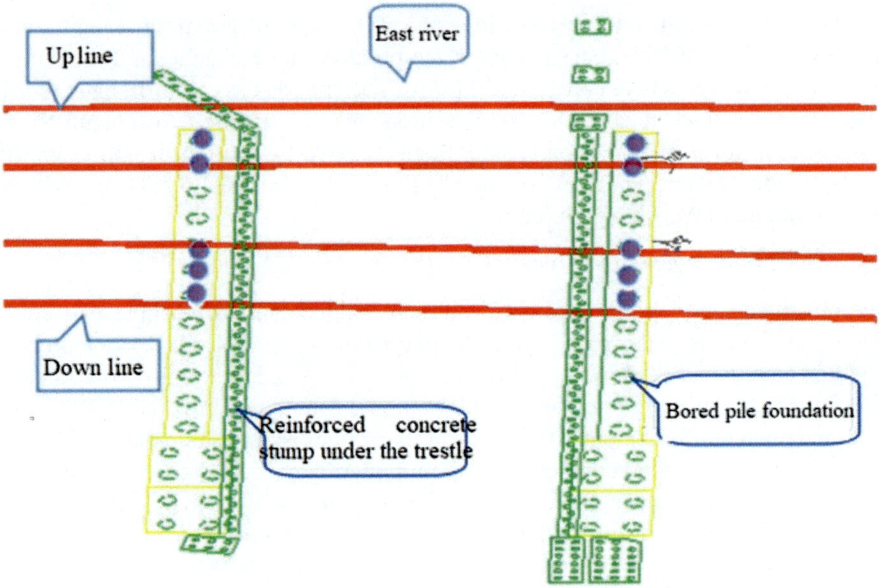

Fig. 8.2 Location plan of the tunnel and Fengqi Bridge

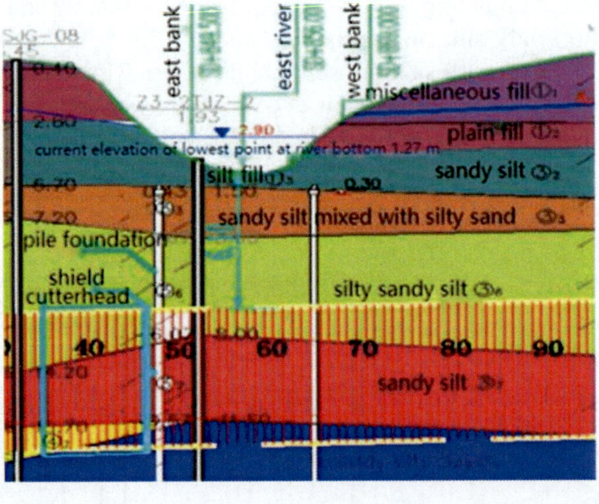

Fig. 8.3 Geological section of Fengqi Bridge and Donghe river crossed by shield

(4) Hydrogeological conditions across the Fengqi Bridge section

During the construction of shield tunneling across Fengqi Bridge, the shield tunneling machine was located below the East River, 2 m deep, and 6.1 m from the tunnel. The groundwater in the strata in the crossing section is mainly porous diving, mainly occurring in the surface fill soil and the and layer of ②, ③ silty soil and silty sand. It

Table 8.1 Description of main stratigraphic characteristics across fengqi Bridge Section

Layer no	The name of the	Geotechnical characteristics
③$_2$	Sandy silt	Grayish yellow, wet, loose to slightly dense. It contains a lot of ferric oxide porphyritic body, mica debris, and has strong viscosity in local area. The shaking reaction was medium, lackluster reaction, low dry strength and low toughness. It is a medium compressible soil
③$_3$	Sandy silt is mixed with silt	Gray, bluish gray, slightly dense–medium dense, wet. Containing mica debris and containing silty sand, the shaking reaction is quick, the luster reaction is low, the dry strength is low, the toughness is low. It is a medium compressible soil
③$_6$	Silty silt with sandy silt	Sallow, bluish grey, very wet, medium dense. It contains iron oxide, mica chips, shell chips, and a small amount of sandy silty and cohesive soil. The shaking reaction is rapid, lackluster reaction, dry strength is low, toughness is low. Belongs to medium low compressibility soil
③$_7$	Sandy silt	Grey, very wet, slightly dense. Containing iron oxide and mica chips, the layer is interbedded with silty masses. The shaking reaction is rapid, lackluster reaction, dry strength is low, toughness is low. It is a medium compressible soil
⑥$_1$	Silty clay	Gray to dark gray, mainly plastic flow. Containing organic matter, locally containing shell debris, high sensitivity. Rough cut surface, no shaking reaction, medium dry strength, medium toughness. It belongs to high compressibility soil

is replenished by atmospheric precipitation runoff. The water volume is medium to large, and the groundwater level changes with the seasons. During the exploration, the buried depth of the borehole static water level was measured to be 2.0 m–2.8 m, and the corresponding elevation was 4.15 m–5.97 m.

8.2.2 Reconstruction and Reinforcement Measures of Shield Tunneling Machine

(1) Reconstruction of Herrick shield machine

1) Reconstruction of knife plate

First kinfe reconstruction: The knife disk of Herrick S-664 shield machine was originally designed for soft soil. In order to adapt to the new geology and grind piles across the river, it needs to be reconstructed again. Based on the geological conditions of the whole shield region, the tool distribution as shown in Fig. 8.4 is adopted.

Table 8.2 Statistical table of physical and mechanical properties of main strata across Fengqi Bridge section

Layer no	Name of the soil	Statistics	A physical indicator of a natural state					
			The water content	The density of		The proportion of soil particles	Void ratio	Saturation
				Wet	Dry			
			W_0	ρ	ρ_d	G_s	e	S_r
			%	g/cm^3		–	–	%
⑥$_1$	Silty clay	The average	35.7	1.85	1.37	2.72	1.015	97.2
③$_2$	Sandy silt	The average	29.0	1.88	1.46	2.70	0.8543	91.7
③$_3$	Sandy silt is mixed with silt	The average	25.1	1.96	1.57	2.70	0.7206	94.0
③$_6$	Silty silt with sandy silt	The average	25.8	1.93	1.54	2.70	0.7565	92.1
③$_7$	Sandy silt	The average	26.7	1.91	1.51	2.70	0.7935	91.3

Table 8.3 Statistical table of permeability coefficient of main strata across Fengqi Bridge Sectio n

Layer no	Name of the soil	Laboratory test permeability coefficient (M/D)		Field pumping test permeability coefficient (M/D)
		K_V	K_H	K
③$_2$	Sandy silt	7.11 e-01	8.25 e-02	E-01 e-02 4.17–6.90
③$_3$	Sandy silt is Mixed with silt	7.53 e-02	9.68 e-02	
③$_6$	Silty silt with Sandy silt	5.89 e-02	7.70 e-02	
③$_7$	Sandy silt	7.81 e-02	9.68 e-02	
⑥$_1$	Silty clay	1.73 e-042	3.02 e-04	

Grinding pile can be divided into two parts, one is the cutting of concrete, the other is the cutting of the steel bar inside concrete. For plain concrete pile or plain concrete pile, the cutter plate of shield tunneling machine can be used to grind the pile successfully. The concrete strength of this shield segment is C25, and the cutter plate after the reconstruction of S664 shield machine can deal with the formation condition below 30 MPa. For the cutting of steel bar, the first knife will be replaced by cutting concrete first knife (Fig. 8.5). Ordinary first knife for tunneling in the argillaceous siltstone has good effect, according to the size on the drawing, first dao's height is 200 mm, but there is no enough this knife for cutting reinforced concrete, a combined gold in the middle of ontology for the knife, when cutting steel, if the position of the

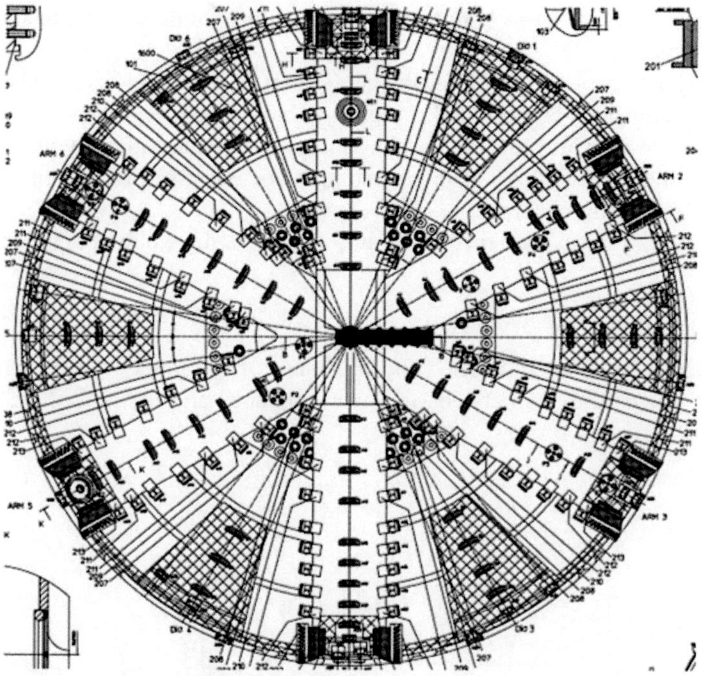

Fig. 8.4 Schematic diagram of knife plate

Fig. 8.5 Cut concrete first knife

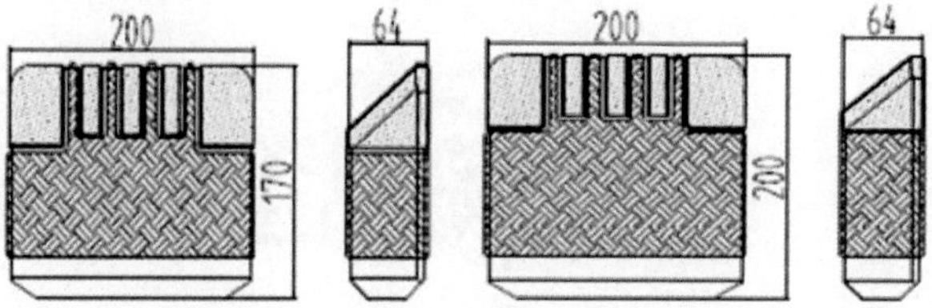

steel bar in the middle of the two pieces of metal that is impossible for steel cutting, even can cause the wear of cutting tool, and affect the alloy block soundness is the final cut as a result of the cutting of reinforcement length is differ, may be wrapped in cutting tools, can also accumulate in the soil bin, even stuck in a screw conveyor. Concrete cutter has a unique design, both sides are big round alloy, the middle is an alloy, to ensure that the whole cutting surface is alloy. The two sides of the concrete cutter are designed with large rounded corners, which can reduce the impact when the cutter rotates. Cut concrete first the top of the knife blade is narrower, less stress, easy to cut, in contact with the steel, more conducive to cutting.

Gear cutter modification: the original gear cutter is light gear cutter, width is only 100 mm, the load is small, impact resistant, easy to cause the situation of falling off.

The new gear cutter is a heavy-duty gear cutter with a width of 250 mm and a better impact resistance.

Transformation of center knife: the center part of the knife disk is also the key part of grinding pile. The original design is fishtail type of center soft earth knife. There is no specific purpose for grinding piles. The center knife layout for cutting concrete is shown in Fig. 8.6. The protruding blade has a similar structure with the first knife for cutting concrete.

Wear-resisting measure: add 3 mm thick wear-resisting welding to the knife plate (Fig. 8.7).

2) Modification of screw machine

By increasing the wear resistance and thickness of the inner wall of spiral blade and spiral sleeve, the gap between the two is reduced and the risk of the steel bar

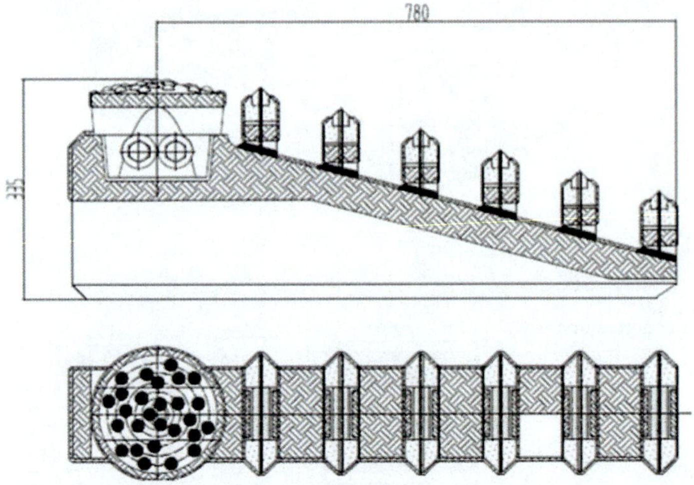

Fig. 8.6 Schematic diagram of center tool arrangement

Fig. 8.7 Abrasion—resistant welding of knife plate

being stuck between the two is reduced. The function of the screw gate is added to ensure the sealing performance of the screw during crossing the East River and pile grinding.

Wear resistance measures: the first three shafts and blade surfaces are all overlaid with wear resistant layers, while the remaining shafts and blade surfaces are overlaid with dense mesh wear resistant layers. The front 3.5 sections of spiral blade are surrounded by wear-resistant alloy blocks; wear-resistant welding of the inner cylinder wall at the front end; Wear resistance welding of 1500 mm pile in front of fixed section inner cylinder wall. Modified spiral blade is in Fig. 8.8 and reformed screw machine is in Fig. 8.9.

Fig. 8.8 Modified spiral blade

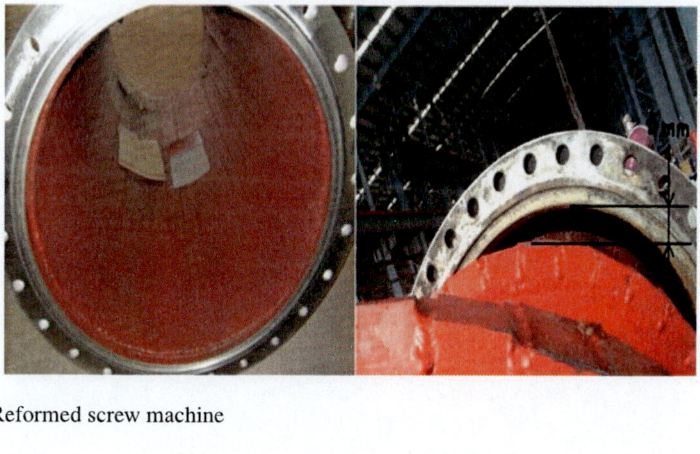

Fig. 8.9 Reformed screw machine

The front and rear gates of the auger guarantee the sealing performance of the auger during crossing (Figs. 8.10 and 8.11).

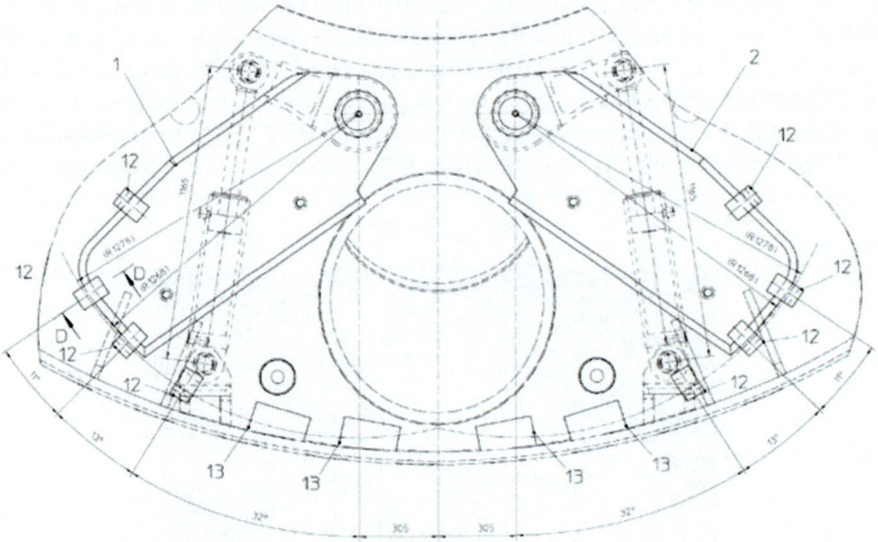

Fig. 8.10 A schematic diagram of the opening state of the screw front gate

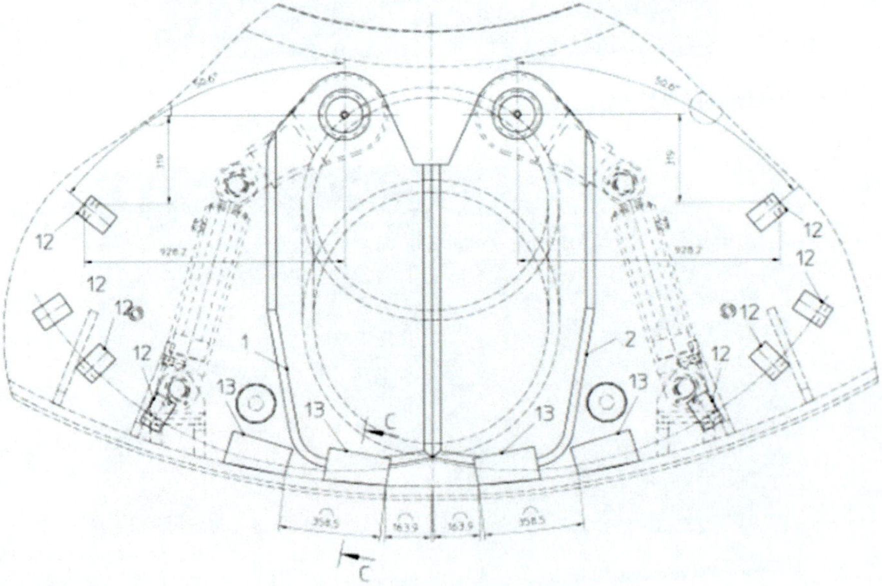

Fig. 8.11 A schematic diagram of the closing state of the screw front gate

3) Low speed pump system

Low speed pump system can be high pressure constant speed propulsion, propulsion speed can be set manually and mechanically, keep the propulsion speed stable, avoid the deviation of electronic control speed regulation. When using the low-speed function, the original propulsion pump will be stopped, and the low-speed pump will be used only for constant speed propulsion. It can also be used as a supplement to the original propulsion pump to improve the propulsion speed.

Due to the existence of steel reinforcement in the pile foundation, in order to protect the tool, the cutter should be "gnawing" a little at a time when the cutter is cutting, that is, "grinding" is the basic principle of pile cutting. However, according to the previous experience, when cutting pile, the cutting resistance and pushing resistance of the cutter are subject to dynamic changes. If the original large-flow jack of shield is used to push the cutter, it will be difficult to control the actual pushing speed within the stable range due to its thick flow range of single adjustment. Therefore, a low-speed and low-flow propulsion pump is added to ensure that the shield tunneling machine can advance at a low speed and a steady speed when cutting the pile foundation, so as to avoid the influence of the speed on the pile machine.

4) The shield-tail brush is strengthened

In the process of tunneling, if the knife plate of the shield tunneling machine gets stuck, the shield tunneling machine must retreat, so that the shield tail brush will be seriously damaged. Therefore, the strengthened shield tail brush is adopted when selecting the shield tail brush, so as to prevent the shield tail brush from serious damage and leakage of mud and water. As shown in Fig. 8.12.

(2) Reinforcement of Fengqi Bridge

Before the shield mill pile, the foundation of fengqi bridge should be reinforced by supporting. Raft foundation should be used to replace the original pile foundation to bear the load brought by the upper part. The main supporting foundation of the bridge in the crossing range of shield structure should be replaced. After the bridge foundation is replaced and strengthened, the pile is ground by shield. Elevation drawing of Fengqi Bridge reinforcement and shield section shows in Fig. 8.13.

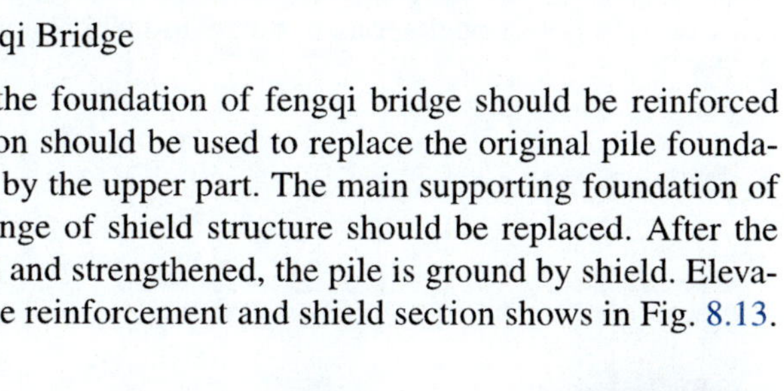

Fig. 8.12 Enhanced shield-tail brush

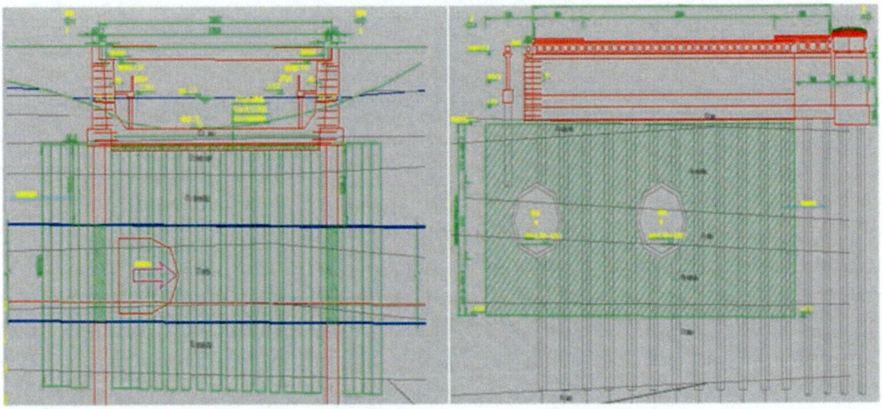

Fig. 8.13 Elevation drawing of Fengqi Bridge reinforcement and shield section

Schematic diagram of raft foundation structure shows in Fig. 8.14. Reconstruction and reinforcement plan of Fengqi Bridge shows in Fig. 8.15.

(3) Improve tunnel strength grade and stability

After the reinforcement, the load of Fengqi Bridge is acted on the raft foundation. But for the long-term consideration of tunnel and bridge operation, ultra-deep lining rings (27 rings for both upper and lower lines) are selected for lining rings of shield tunneling through Fengqi Bridge section (length SDK21 + 838.00–SDK21 + 870.00, XDK21 + 836.00–XDK21 + 868.00), and the strength of connecting bolts is of grade 8.8. Fig. 8.16 is section diagram of the relationship between shield tunnel and Fengqi Bridge.

Fig. 8.14 Schematic diagram of raft foundation structure

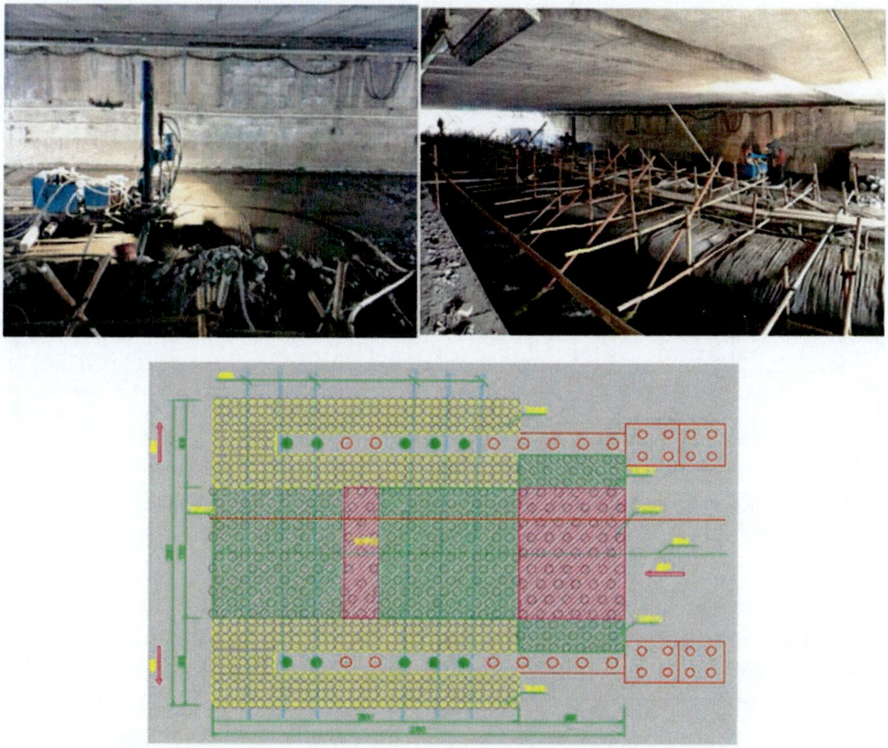

Fig. 8.15 Reconstruction and reinforcement plan of Fengqi Bridge

8.2.3 Analysis of Control Technology for Pile Grinding Across River

(1) Tunneling time

Pile foundation of fengqi Bridge in the first row:

Starting and ending time of the 43rd ring heading: 2016.1.11 07:55–13:10;

The start and end time of the 44th ring heading: 2016.1.11 14:10–22:10.

Pile foundation of fengqi Bridge in the second row:

Start and end time of the 60th Ring heading: 02:35–09:02, 2016.1.15;

Start and end time of 61st ring heading: 10:33–16:15, January 15, 2016.

On January 16, 2016, the shield tail detached from the pile foundation of Fengqi Bridge.

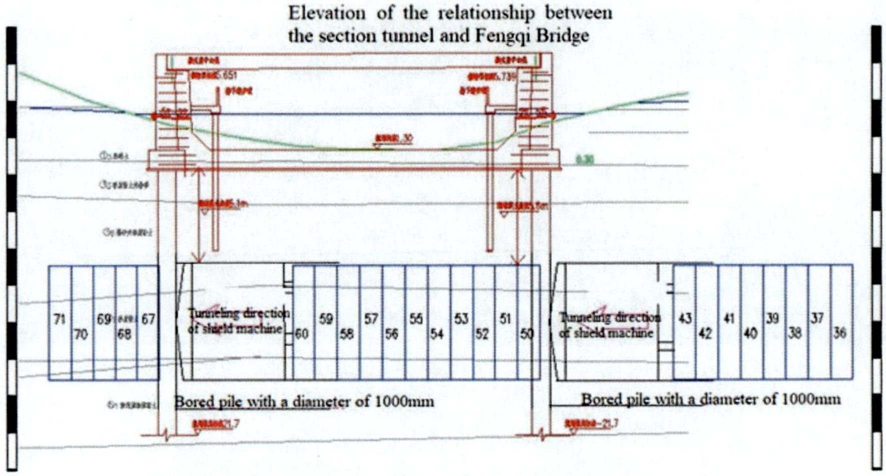

Fig. 8.16 Section diagram of the relationship between shield tunnel and Fengqi Bridge

(2) Thrust

During the construction of the first row of pile foundation grinding pile, the thrust is 1400 t–1450 t. After passing through the pile foundation, the thrust gradually decreases, and the late driving thrust of the 44th ring falls to 1200 t. During the construction of grinding pile in the second row of pile foundation, the thrust is 1100 t–1150 t. After passing through the pile foundation, the thrust decreases, and the driving thrust of the 62nd ring falls to 1000 t.

(3) Driving speed

The constant-speed pump is started from one ring before pile grinding construction to one ring after pile grinding construction, and the constant-speed driving is controlled at 5 mm/min. The driving speed of the shield cutter plate increases gradually after it passes through the whole cutter plate.

(4) Rotary speed of knife plate

The rotary speed of the cutter plate was controlled at 0.8 rpm/min during the whole pile grinding process.

(5) Grouting pressure and grouting volume

The grouting pressure is 4–5 bar, and the grouting volume is $4m^3$.

(6) Earth pressure

The No. 1 soil pressure was controlled at 1.5 bar during the whole pile grinding process.

Fig. 8.17 Photos of concrete and reinforcement taken out during pile grinding

(7) Torque

First row pile foundation: 2.0 MN·m–2.2 MN·m; the second row pile foundation: 2.0 MN·m–2.6 MN·m.

(8) Analysis of pile shield parameters

In the process of pile grinding construction, constant speed pump is used, so that the driving speed is well controlled and the driving speed is slow, so that the main reinforcement of pile foundation is well cut, the main reinforcement is basically broken, and the risk of steel reinforcement winding knife disk is reduced. According to the existing construction cases in China, the best rotation speed was determined to be 0.8 rpm /min. The setting of earth pressure is calculated with safety factor fully considered. The grouting pressure and grouting quantity are determined according to the early driving parameters and settlement.

When the shield grinding pile passes through the first row of pile foundation, the thrust is increased due to the large hinge pressure of the shield tail, and the hinge pressure is reduced by adjusting before the grinding pile passes through the second row of pile foundation, so the thrust is reduced to some extent when the grinding pile passes through the second row of pile foundation.

The torque variation of cutter plate is small in the whole process of pile grinding. When the grinding pile passes through the second row of pile foundation, the torque of the cutter head is reduced from 2.6MN·m to 2.2MN·m by improving the residue in front of the cutter head. Photos of concrete and reinforcement taken out during pile grinding is in Fig. 8.17.

(9) Summary of monitoring data

Figure 8.18shows the monitoring site layout of pile grinding crossing area. After grinding the first row of piles, the maximum cumulative settlement of the bridge deck is −0.97 mm (QCJ11), while after grinding the second row of piles, the maximum cumulative settlement of the bridge deck is −3.09 mm (QCJ2). After grinding the first row of piles, the maximum cumulative settlement of the piers is 1.49 mm (QC2). After grinding the second row of piles, the maximum cumulative settlement of the piers is −2.38 mm (QC4) .

Based on the above data analysis, it can be concluded that the upper line shield pile grinding crossing interval, and the cumulative settlement of the bridge deck is only −3.09 mm, without single alarm or cumulative alarm. Generally speaking, the upper line shield driving parameters are well controlled, without significant influence on The Fengqi Bridge and the river. Figures 8.19, 8.20 and 8.21 are the settlement process curves.

.

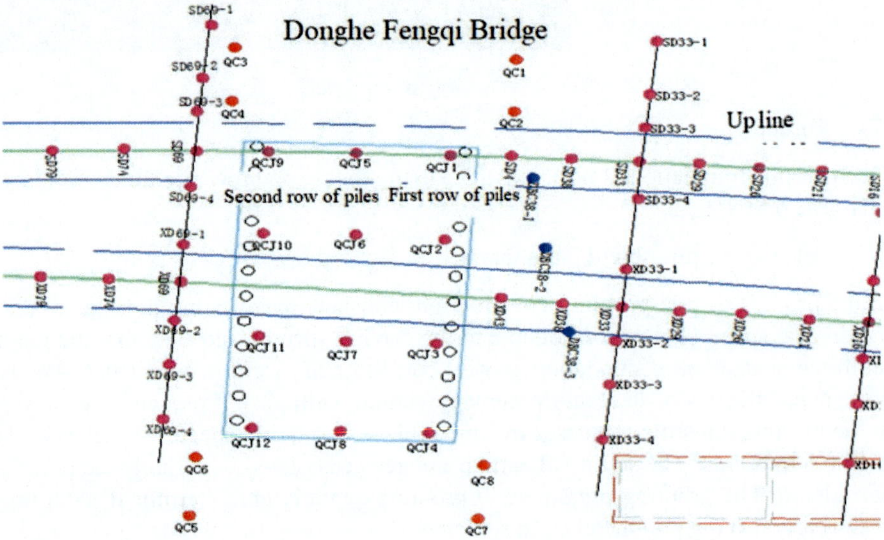

Fig. 8.18 Monitoring site layout of pile grinding crossing area

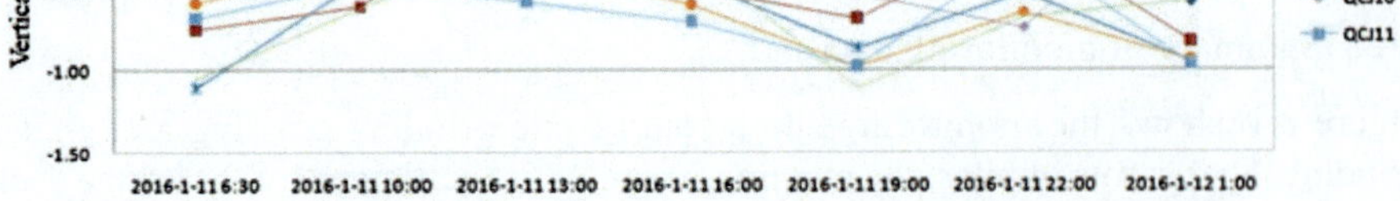

Fig. 8.19 Bridge deck settlement process curve during grinding the first row of piles

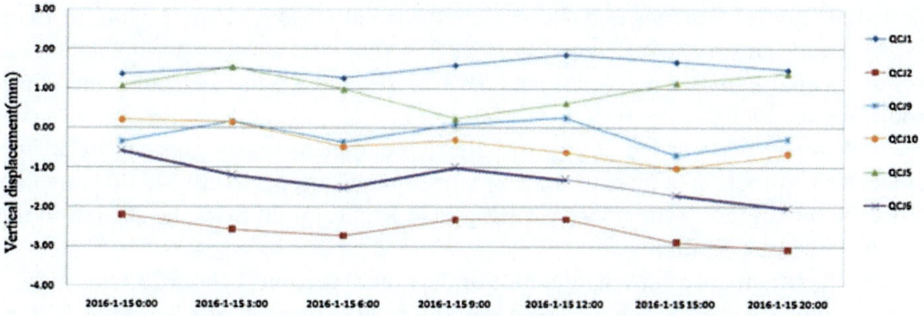

Fig. 8.20 Curve of bridge deck settlement process during grinding the first row of piles

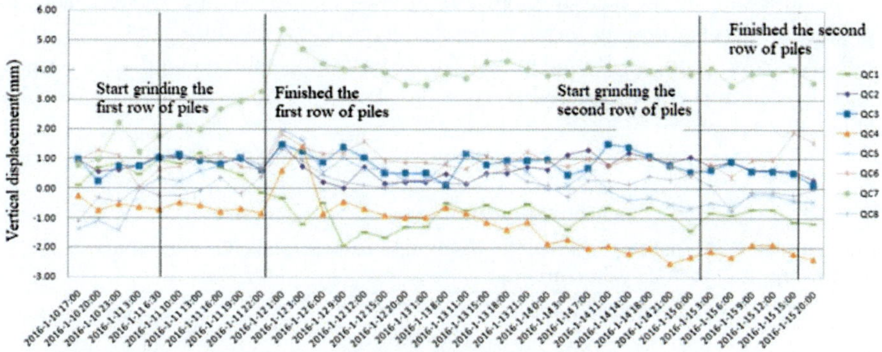

Fig. 8.21 Settlement process curve of abutment during pile grinding crossing

8.3 Case 2: The Tunnel Between Jiaoqiao Station and Changjiang Road Station in Nanchang Undergoes the West Part of Fenghuang Garden

8.3.1 Project Overview

(1) Project overview

The second civil engineering standard of the first phase of Nanchang Rail Transit Line 1, the tunnel project between Jiaoqiao Station and Changjiang Road Station is located in Honggutan New District, Nanchang City. Tunnel from the dumpling bridge station after south jiangxi traffic stadium near the southern tip of vocational and technical college, a loop line across the north to the south, underpassing Yingshang Ying Lake, along the dimness in phoenix garden residential area 340 m radius curve right connected feng and north road, access road station, interval od range of SK1 + 570.427–SK4 + 466.174(set a middle wind well in the middle), it's 2895.747 m long, downlink 2895.399 m long and interval length is totally 5791.146 m. An intermediate air well and liaison channel is set up at SK3 + 098.821, another liaison channel is set up at SK2 + 050, SK2 + 500 and SK3 + 670 and a liaison channel and pump room are set up at SK4 + 035.

The project was carried out by two Herrick composite earth pressure balancing shields. Construction of up-line and down-line tunnels successively. Shield tunneling from the head shaft in the north of Changjiang Road station to the head shaft in the south of Jiaoqiao Station.

The maximum overburden thickness of the tunnel section is about 23.6 m, and the minimum overburden is 4.9 m. The minimum radius of the interval curve is 340 m, and the line spacing is about 12–13.6 m. The longitudinal slope of the line is designed as a two-way slope with a maximum slope of 29‰.

The interval should pass through J-01#, J-2# and GJ04, the western area of Fenghuang Garden, mainly through the soil layer of $②_4$ medium sand, $②_5$ Coarse sand, $②_6$ Gravel sand, $②_7$ Round gravel, $⑤_{1-1}$ Strongly weathered argillaceous sandstone, $⑥_1$ Completely weathered phyllite, $⑥_2$ Highly weathered phyllite, etc (See Table 8.4).

Table 8.4 Analysis table of tunnel structure geology

The construction of the methods	Head height (From bottom plate) (m)		Rock and soil layer within the interval tunnel		
	Originating	Arrive	Name of the geotechnical	Set period of	The mileage
Composite earth pressure balance shield	10.24	12.28	$⑥_1$ Completely weathered phyllite, $⑥_2$ Strong weathering of phyllite, $⑤_{1-1}$ Highly weathered argillaceous siltstone	F	CK1–CK3 + 507.438 + 570.423)
			$②_7$ Round gravel, $⑤_{1-1}$ Highly weathered argillaceous siltstone, $⑤_{1-2}$ Moderately weathered argillaceous siltstone	F	CK3–CK4 + 276.622 + 507.438)
			$②_4$ Medium sand, $②_5$ Coarse sand, $②_6$ Gravel sand, $②_7$ Round gravel	E	CK4–CK4 + 466.094 + 276.622)

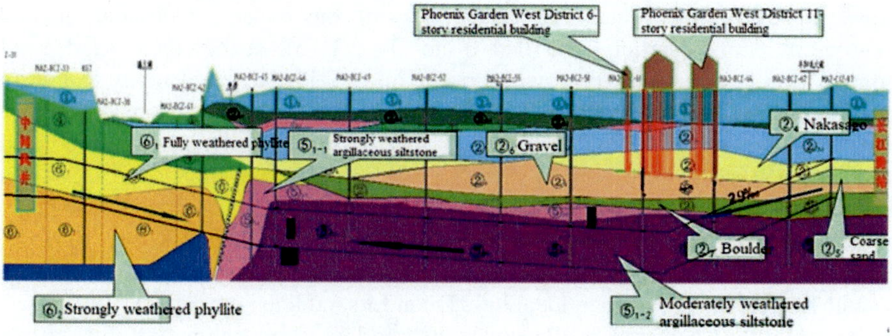

Fig. 8.22 Schematic diagram of building location strata under construction

(2) Geological conditions

The interval tunnel passes through the stratum is mainly composed of ②$_4$ Medium sand, ②$_5$ Coarse sand, ②$_6$ Gravel sand, ②$_7$ Round gravel, ⑤$_{1-1}$ highly weathered argillaceous sandstone with complex geological conditions and great difficulty in construction (Fig. 8.22).

The stratigraphic distribution of the site is described in detail from top to bottom as follows:

Engineering Geology SECTION II

①$_2$ Plain fill: loose, gray, grayish yellow, mainly composed of sandy soil, locally composed of cohesive soil, containing a small amount of gravel, mixed lithology, distribution of exploration holes on the subgrade, revealing that the upper 50–70 cm is dominated by gravel.

②$_2$ Silty clay: fluid plastic, high compressibility, gray, layered, layer with thin layer of silty sand, single layer of silty sand 0.1–0.4 cm thick, local lumpy silty sand, containing a small amount of humus, no shaking reaction, smooth section, dry strength, moderate toughness.

②$_{3-1}$ Cohesive soil silty sand: moderately high compressibility, looseness, grayish yellow, local pore segments are brown yellow, with plasmolytic masses of cohesive soil, mainly composed of quartz, mica and feldspar.

②$_{3-2}$ Fine sand: saturated, moderately compressible, loosely—slightly dense, gray, grayish yellow, part of the pore section is medium sand, local contains a small amount of argillaceous, the composition is mainly quartz, mica, feldspar.

②$_4$ Medium sand: saturated, medium compressibility, slightly dense-medium density, gray and grayish yellow, part of pore segments are fine sand, mainly composed of quartz, mica and feldspar, and locally containing a small amount of gravel, mainly with gravel diameter <1 cm.

②$_5$ Coarse sand: saturated, medium compressibility, medium density, gray, gray yellow, gray white, mainly composed of quartz, mica, feldspar, containing gravel, gravel diameter <1 cm mainly, content of about 5–10%.

②$_6$ Gravel sand: saturated, medium compressibility, slightly—Micronesia, grey, gray, particle size is given priority to with 0.2–2 cm, 20–35%, parent rock composition is given priority to with quartzite, sandstone, siliceous rocks, and circular, drilling exposes maximum particle size of 4–6 cm, 5–15%, fill in the coarse sand, sand composition is given priority to with quartz, mica, feldspar, accidentally caught in the thin layer of coarse sand lens.

②$_7$ Round gravel: saturated, medium low compressibility, Micronesia, local slightly thick, gray, gray, particle size is given priority to with 1–3 cm, 35–45%, content of parent rock composition is given priority to with quartzite, sandstone, siliceous rocks, and circular, drilling exposes maximum size 5–8 cm, 10–15%, content of local stone, filling in the coarse sand, sand composition is given priority to with quartz, mica, feldspar, local hole Duan Li sand lens.

⑤$_{1-1}$ Highly weathered argillaceous siltstone layer: low compressibility, purple red, the original rock structure is visible, the core is relatively broken, in the shape of fragments or short columns, the fragments can be broken by hand, softening in water, normal drilling speed is faster, the core recovery rate is low.

⑤$_{1-2}$ Weathering of argillaceous siltstone, amaranth, partial section of the celadon, argillaceous cementation, local for calcium cementation, the original rock structure is clear, core is complete, a columnar—long column is given priority to, partial to fragmental, joint fissure development, see a little steep dip Angle fractures, hammering sound slightly brittle and break easily, meet water softening, easy disintegration plate shaped.

(3) General situation of the crossing building (structure)

This section has successively passed through GJ04, J-02# and J-01# of the West district of Phoenix Garden, and finally passed through the side of G02 of the West District of Phoenix Garden. For details, please refer to the plan figure of the relationship between the shield section and fenghuang Garden community (Figs. 8.23 and 8.24).

1) Phoenix West District Public Construction-GJ04

GJ04 is a 2-storey and partially 3-storey brick concrete structure in the West district of Phoenix. The foundation is reinforced by cement mixing pile (fully reinforcement) with a pile length of about 8.5 m. The buried depth of the tunnel centerline is about 19.5 m, and the top of the tunnel is about 11 m from the pile bottom. When the shield tunneling is carried out on ring 171 of the upper line and ring 194 of the lower line, the cutter has begun to enter the lower part of the building. When the shield tunneling is carried out on ring 205 of the upper line and ring 206 of the lower line, the shield tail passes through the building. The tunnel is located in layer ②$_6$ Gravel sand, ②$_7$ Round gravel, ⑤$_{1-1}$ Highly weathered argillaceous siltstone.

2) West area of Phoenix Garden, J-02

The West district of Phoenix Garden J-02# is an 11-story reinforced concrete short-limb shear wall structure with a design life of 50 years, a total building height of 34.8 m and a design safety level of grade II. The foundation reinforcement was tamped and expanded cast-in-place pile with a length of 13.85 m and a bottom elevation of 3.4 m. The vertical curve of the tunnel axis is 29‰ slope, the minimum distance

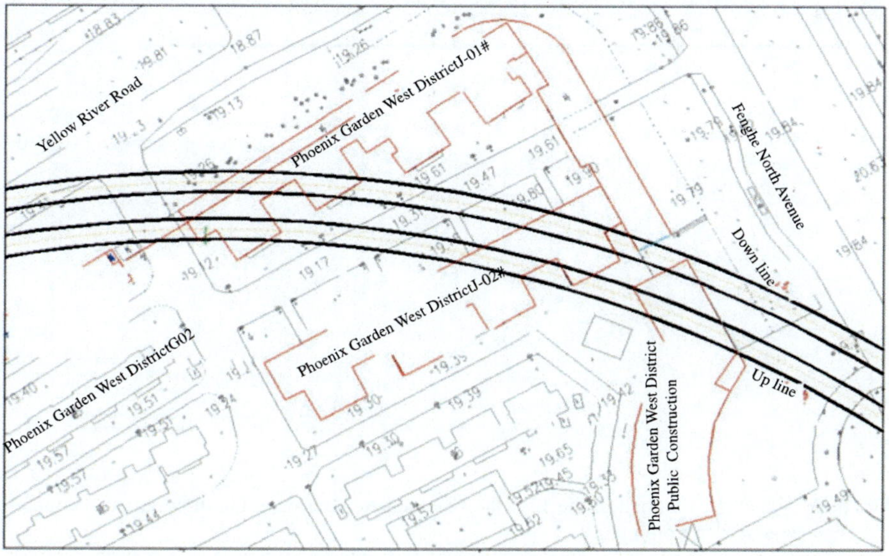

Fig. 8.23 The plan of the relationship of fenghuang Garden community

Fig. 8.24 A schematic diagram of the current shield position of the downline

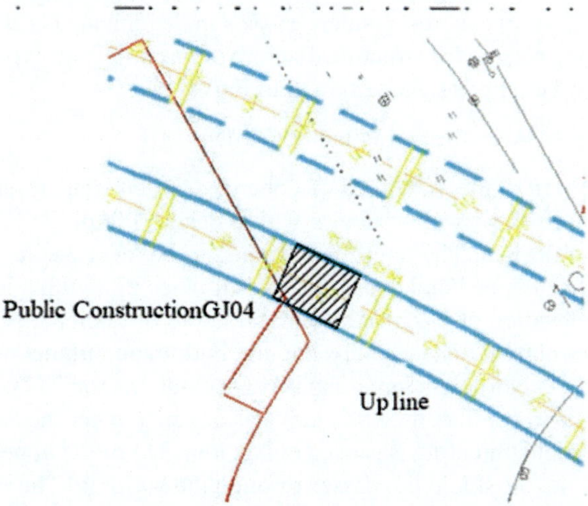

between the pile bottom and the outside diameter top of the tunnel is 67 cm, and the distance from the outside diameter of the shield tunneling machine is 49 cm.

By the time shield tunneling reaches loop 209 of the upper line and Loop 207 of the lower line, the knife disk has begun to enter the lower part of the building. By loop 246 of the upper line and 242 of the lower line, the shield tail passes through the building. The upper line crosses about 24 pile foundations, and the lower line crosses about 26 pile foundations. The tunnel is located in layer ②$_6$ Gravel sand, ②$_7$

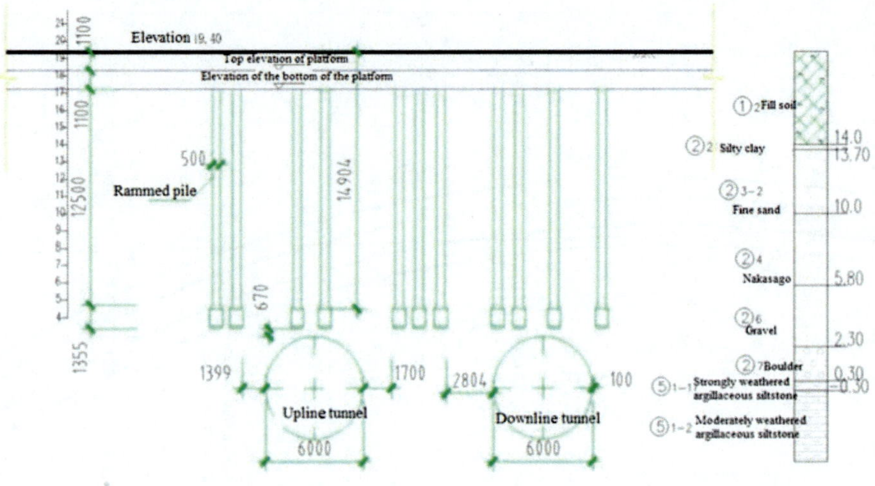

Fig. 8.25 Schematic diagram of the 11th-floor residence section running through J-02# concrete in Fenghuang Garden Community

Round gravel, ⑤$_{1-1}$ Highly weathered argillaceous siltstone layer, ⑤$_{1-2}$ Moderately weathered argillaceous siltstone. For details, please refer to the section diagram of 11-story concrete residence j-02 # in Fenghuang Garden Residential area in Fig. 8.25, and refer to the structure diagram of ramming and expanding pile of 11-story building in Fenghuang Garden 8–26 in Fig. 8.26.

3) West Area of Phoenix Garden, J-01

J-01# in the West area of Phoenix Garden is an 11-story reinforced concrete short-limb shear wall structure with a design life of 50 years and a total building height of 36.1 m. The house is designed to be level 2 safe. The foundation reinforcement was tamped and expanded cast-in-place pile with a length of 13.85 m and a bottom elevation of 3.4 m. The vertical curve of tunnel axis is 29‰ vertical slope, and the minimum distance between pile bottom and tunnel outer diameter top is 3.45 m.

When shield tunneling is carried out on loop 277 of the upper line and loop 276 of the lower line, the knife disk has begun to enter the lower part of the building. When shield tunneling is carried out on loop 311 of the upper line and loop 316 of the lower line, the shield tail passes through the building. The upper line passes through about 31 pile foundations, and the lower line passes through about 40 pile foundations. The tunnel is located in layer ⑤$_{1-1}$ Highly weathered argillaceous siltstone layer, ⑤$_{1-2}$ Moderately weathered argillaceous siltstone. For details, see Fig. 8.27 for the section diagram of 11-storey concrete residence J-01 in Fenghuang Garden Community.

4) West area of Phoenix Garden, G02

G02 in the west area of Phoenix Garden has 6 floors and 7 floors, with a total building height of 23.05 m. The design life is 50 years, and the structural design safety level is level 2.1 floor is frame structure, 2–7 floors are brick and concrete structure. The

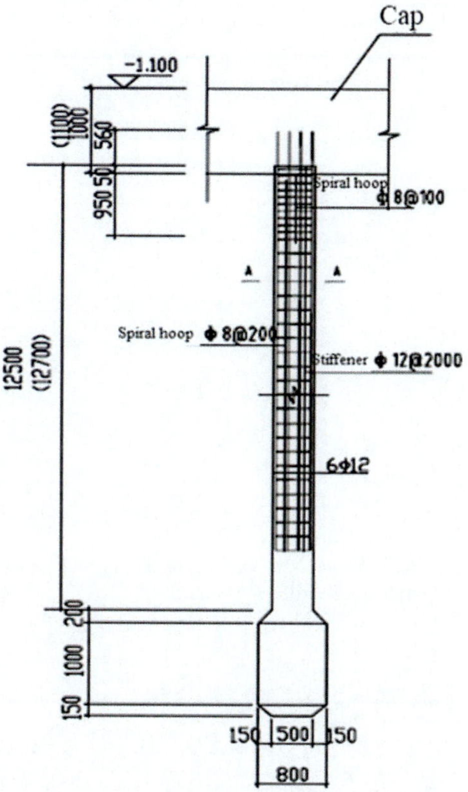

Fig. 8.26 Structure drawing of rammer expanded pile for 11-story building in Fenghuang Garden

foundation reinforcement was tamped and expanded cast-in-place pile, the length of which was about 12.35 m, and the elevation of the bottom of the pile was 6.05 m. The shield tunneling axis passes beside it. The distance between the center line of the tunnel and the center of the peripheral pile is 5.4 m, and the vertical distance between the pile bottom and the top of the tunnel is 6.85 m. The tunnel is located in layer ⑤$_{1-1}$ Highly weathered argillaceous siltstone layer, ⑤$_{1-2}$ Moderately weathered argillaceous siltstone. See Fig. 8.28 for details. Section diagram of 6-storey concrete residence in Fenghuang Garden Community.

(5) Conclusion of housing appraisal report

According to the appraisal, no cracks were found in G02, J-01#, J-02# and GJ04 (2–34)–(3–8) shaft beams, slabs and columns of Fenghuang Garden Public Construction In the West area of Fenghuang Garden. The main structure is safe, and cracks and water seepage in some parapet walls do not affect the safe use of the main structure.

According to the relevant design specifications and building investigation situation, as well as the housing appraisal report, comprehensive analysis and research to determine the building deformation control standard, guide the construction, to ensure the building safety. The criteria are as follows:

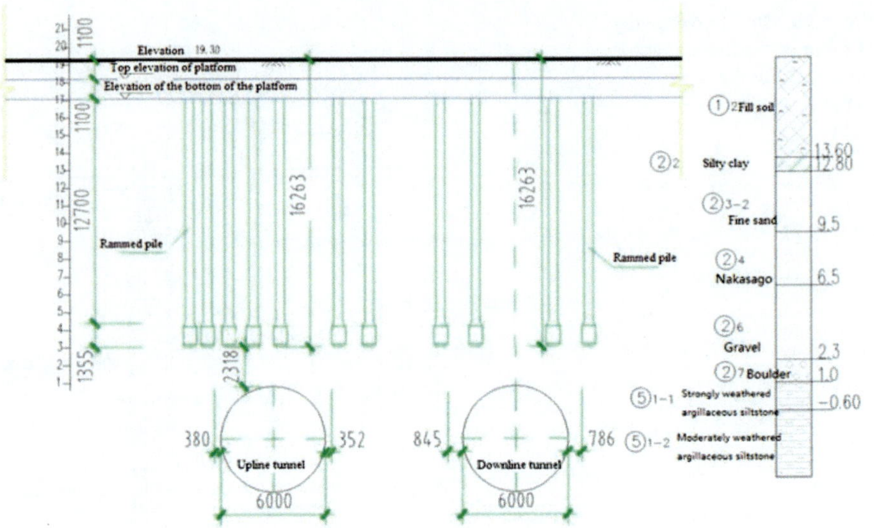

Fig. 8.27 Schematic diagram of the 11th-floor residence section running through J-01# concrete in Fenghuang Garden Community

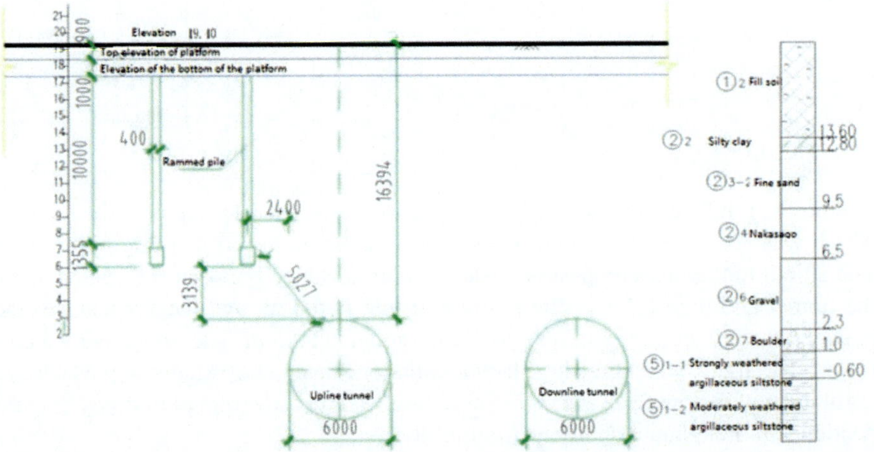

Fig. 8.28 Section diagram of 6-storey concrete house passing through Fenghuang Garden Area

The relative tilt of the building (structure) is not more than 2‰;

The horizontal displacement of pile foundation shall be no more than 5 mm and the vertical displacement shall be no more than 10 mm.

The surface subsidence shall not be greater than 30 mm, and the surface uplift shall not be greater than 10 mm.

8.3.2 Grouting Reinforcement and Emergency Plan

According to the special Design of the section from Jiaoqiao Station to Changjiang Road Station of the first Phase of Nanchang Rail Transit Line 1 crossing the West District of Fenghuang Garden, measures such as embedding sleeve valve pipe and increasing the amount of secondary grouting are taken for shield tunneling through the building.

(1) Situation analysis

The buried depth of the tunnel is about 16 m, the vertical curve of the tunnel axis is 29‰, and the minimum distance between the pile bottom and the outside diameter top of the tunnel is 1.24 m. Because the tunnel is buried deep, some of it is located in the medium weathered argillaceous siltstone, and the shield is close to the pile foundation, if the soil around the interval tunnel is reinforced, the effect of reducing the influence of shield construction on the soil and pile foundation around the tunnel is little effective, so the soil around the shield tunnel is not reinforced in this construction. However, in order to deal with the large deformation and settlement of the surface buildings caused by improper operation during construction, sleeve valve tube embedded in the peripheral tunnel crossing in the j-02 residential building (Fig. 8.29) was followed by the sleeve valve tube grouting according to the monitoring situation of shield construction, so as to ensure the shield construction to cross the building safely and stably.

The length of the upper line of the interval tunnel is SK4 + 69.773–SK4 + 135.773, and the length of the lower line is XK4 + 82.626–XK4 + 148.626, which

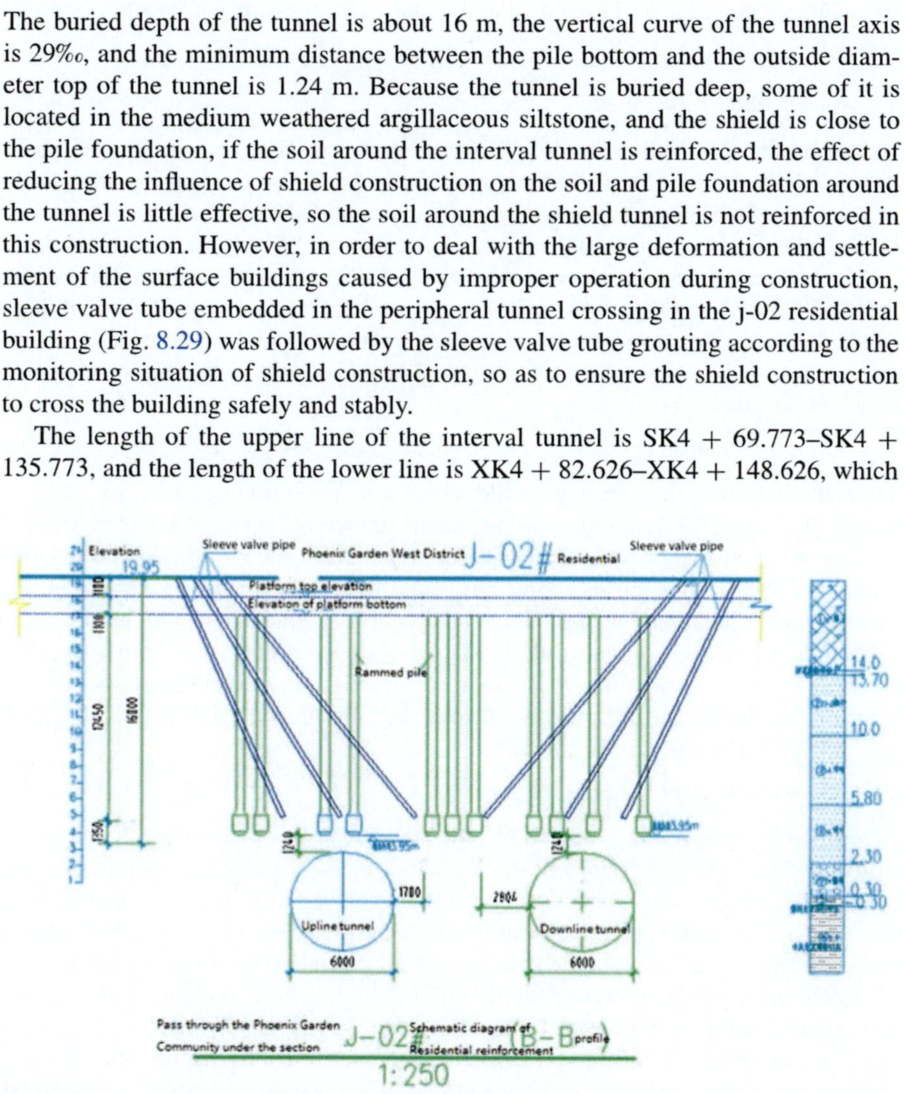

Fig. 8.29 Schematic diagram of sleeve valve pipe layout

goes through the J-01 residential building in the West area of Fenghuang Garden. The stratum of this interval tunnel is medium weathered argillaceous siltstone. The buried depth of the tunnel is about 19 m, the vertical curve of the tunnel axis is 29‰, and the minimum distance between the pile bottom and the outside diameter top of the tunnel is 3.45 m. Considering that the tunnel is deeply buried and the surrounding rocks are hard to pass through, the construction technology and quality should be controlled well during the construction, the driving parameters of shield tunneling should be optimized, and the monitoring should be strengthened. Therefore, shield tunneling has little influence on the crossing construction, so no pre-reinforcement measures should be taken for the J-01# residential building in the west area of Phoenix Garden.

(2) Pipe range and grouting control of sleeve embedded valve

1) Sleeve valve pipe range

Before the shield tunneling through the building, sleeve valve pipe construction shall be carried out. During the shield tunneling, sleeve valve pipe tracking grouting shall be carried out according to the monitoring situation of shield tunneling. This time, sleeve valve pipe construction is mainly carried out for J-02# residential building.

The embedded range of sleeve valve pipe is 3 m in each side of the tunnel where the shield tunneling through the building area and 3 m in each side of the building. Three rows are arranged in a plum—shaped arrangement, and the number of sleeve valve tubes is about 400.Specific layout shall be carried out according to the actual situation on site.

2) Grouting control

Interval shield tunneling through buildings is a major source of risks. In order to ensure the safety of buildings during shield tunneling, according to the special safety construction plan and relevant design contents of the crossing buildings, the embedded sleeve valve pipe is adopted in J-02 building to carry out tracking grouting during shield tunneling. During shield construction, monitoring should be strengthened. According to the monitoring results, analysis should be carried out every day to adjust the relevant parameters of shield construction. In case of continuous settlement or settlement exceeding the monitored control value, targeted sleeve valve pipe grouting will be immediately adopted.

3) Sleeve valve pipe grouting parameters

Sleeve valve pipe grouting method has the advantages of low pressure and control, uniform slurry diffusion and repeated injection of the same grouting pipe. At the same time, it is also necessary to consider that grouting reinforcement construction cannot have an impact on the foundation of the building (such as: too high rotary jet grouting pressure may affect the stability of the foundation and other factors).

Sleeve valve pipe is set from both sides of the building to the middle, and the sleeve valve pipe is set to the top of the foundation ramming pile head of the building, with an Angle of about 50°–60° and a length of about 15 m–18 m. If pile foundation is encountered during drilling construction, the hole position construction should be terminated and another borehole should be constructed nearby.

(3) Reinforcing the measures of secondary grouting in the cave by passing through the construction (structure)

According to the special Design of the section between Jiaoqiao Station and Changjiang Road Station of the first Phase of Nanchang Rail Transit Line 1 crossing the West district of Fenghuang Garden, the measures of strengthening the secondary grouting and pipe section for shield tunneling through the building are taken.

Strengthen the scope of secondary grouting in the hole. Considering the influence of the superstructure load on the interval tunnel and the need of shield tunneling through the building to strengthen the secondary grouting, the interval tunnel passes through the building with the upper line length SK4 + 69.773–SK4 + 135.773 and the lower line length XK4 + 82.626–XK4 + 148.626. The interval tunnel passes through the J-01 residential building in the west area of Fenghuang Garden, and the grouting hole is increased from the original 6 holes to 16 holes. The interval tunnel goes through the J-02 residential building in the West district of Phoenix Garden. The amount of reinforcement and grouting holes are added to the interval pieces, and the grouting holes are increased from the original 6 holes to 16 holes.

During the crossing of the building, secondary grouting will be carried out from the 10th ring after the shield tail, and in the future, secondary grouting will be carried out in every fifth ring of driving, with the grouting amount of 2–4 m^3. During this period, the settlement observation will be strengthened, and the secondary grouting will be strengthened according to the settlement situation until the crossing of the building is smooth.

(4) Emergency plan

1) Start-up conditions of emergency plan

The deformation of J01# and J02# buildings was obtained by means of monitoring and the corresponding emergency plan was started. Specific monitoring control values are as follows:

① Surface settlement $\leq$ 30 mm; Surface uplift $\leq$ 10 mm; ② Building tilt $\leq$ building; ③ When the tunnel construction is pushed through the hole diameter of one time, the displacement rate is $\leq$ 5 mm/d; ④ the monitoring and early warning value is 70% of the control value (see monitoring plan for details).

2) Emergency procedures

As for the settlement and tilt of J-02# building, when the monitoring data exceeds the specified value, ground grouting shall be started immediately and the secondary grouting of pipe segments in the hole shall be strengthened:

① In case of an alarm at a measuring point, three rows of sleeve valve pipes embedded in the measuring point before and after the measuring point and within the range of 3 m or so shall be used for grouting, and the grouting pressure shall be controlled at 0.5–1.5 MPa; and fill grouting according to the monitoring situation; The grouting depth starts from the bottom of the sleeve valve tube.

② Secondary grouting should be carried out at 2 rings around the measuring point of the hole door. The secondary grouting quantity should be controlled at 2–4 m times and the grouting pressure should be controlled at about 0.3 MPa. Since the special pieces with additional grouting holes are used for the pipe pieces, it is advisable to use the jumping hole grouting. Strengthen monitoring, according to the monitoring situation, timely adjust the grouting volume and grouting pressure.

For the settlement and tilt of J01# building, when the monitoring data exceeds the specified value, the secondary grouting of the pipe segment in the hole shall be strengthened immediately:

① Secondary grouting shall be carried out at 2 rings around the measuring point of the portal, the secondary grouting quantity shall be controlled at 2–4 m times, and the grouting pressure shall be controlled at about 0.3 MPa; because the special pieces with additional grouting holes are used for the pipe pieces, it is easy to use the jumping hole grouting. Strengthen monitoring, according to the monitoring situation, timely adjust the grouting volume and grouting pressure.

② When the monitoring data meets the specified requirements during the shield tunneling machine crossing the building, no large deformation occurs. During the construction of the shield tunneling machine, synchronous grouting is needed to be increased in the hole to supplement the second grouting, and the sleeve valve pipe is filled with grouting on the ground.

8.3.3 Analysis of Construction Control

(1) Analysis of monitoring data

Dumpling–long range up line since from Wei Dang range most surface subsidence accumulative total value than the control values, the surface settlement process line for nearly a week, as shown in Fig. 8.30. On February 28, up line assembled to 162 ring, cutter head driving to 167 ring position, the surface settlement points behind the shield tail R155, R160 sedimentation volume were greater than 32 mm, shield body R165 point above the settlement amount to 15.7 mm, the settlement of measuring point R170 ahead of the cutter did not exceed the control values; on March 1, the upper line was assembled to ring 164, and the cutter head was driven to ring 169. The surface settlement points behind the shield tail presented a settlement trend, among which the settlement rates of Measuring points R160 and R165 reached 18.6 mm/d and 48.1 mm/d respectively. The cumulative settlement amount of measuring points R170 above the cutter head reached 30.9 mm. On March 2, + 165 ring pieces came out of the shield tail, and the palm face was driven from 169 to 171 rings. The settlement rate of the measuring point R165 reached −21.4 mm/d, and the settlement rate of R170 reached −23.4 mm/d. On March 3, + 167 ring pieces came out of the shield

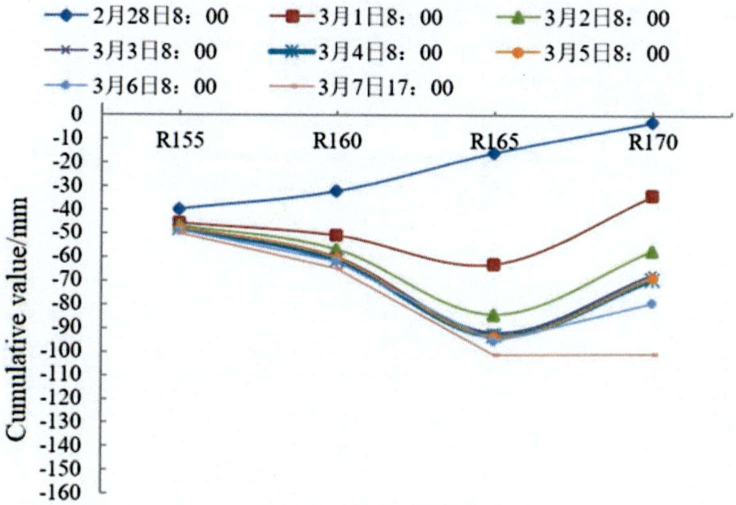

Fig. 8.30 One-week variation process line of surface settlement monitoring points

tail and the palm face tunnelled to 172 rings. The settlement rate of all measuring points decreased, and the maximum settlement rate was −9.9 mm/d (R170). On March 4, only 1 ring was drived in 24 h on the upper line, and the settlement rate at each measuring point of R155–R170 was small. After the shield tunnelling resumed in the afternoon of March 5 and resumed at 17:00 on March 6, the subsidence rate of the measuring point R170 of the maximum subsidence rate on the palm surface reached −28.2 mm/d from the 173-ring tunnelling to the 176-ring tunnelling. During this process, the measuring point R170 gradually moved out of the shield from the top of the shield body. Data of surface settlement monitoring points is in Table 8.5.

See Table 8.6 and Fig. 8.31 for the settlement monitoring data of the two-story commercial building below the shield upline, it can be clearly found that until the afternoon of March 5th, the building settlement monitoring data has been relatively stable with a small settlement trend. After penetrating the building from the palm surface, the settlement trend of each measuring point J01–J06 increased. On March 6, the settlement rate of the measuring point J03 reached 11.9 mm/d, as shown in Fig. 8.32. The measuring point J03 was located below the position of the building GJ04 under the axis of the upper line.

(2) Analysis of shield construction control parameters

After the shield tunneling to ring 150, the site has begun to gradually stabilize the construction parameters of the shield, and prepare in advance for crossing the west area of Phoenix Garden. As shown in Table 8.7, the soil pressure on the site before crossing the building has been raised to 2.0Bar, and the statistical excavated quantity has been stabilized at 38–39m ^{3}, the synchronous grouting volume is stable at 9 m^3; Meanwhile, the total thrust rises to over 15000 kN accordingly, and the cutter torque rises to 2.7×10^3 kN/m, the difficulty of tunneling increases, and the tunneling

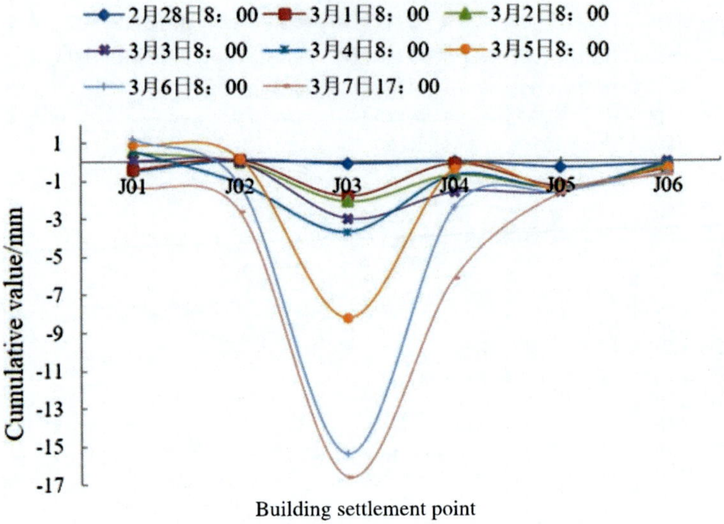

Fig. 8.31 One-week variation process line of settlement monitoring point J01–J06 of two-story commercial building

Fig. 8.32 GJ04 building settlement monitoring point layout

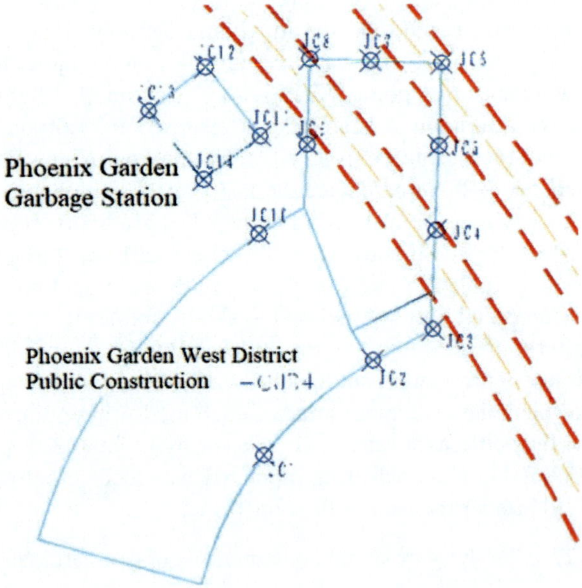

Table 8.5 Data of surface settlement monitoring points (unit mm)

Measuring point day period	February 28th 8:00	March 1 at 8:00	March 2 at 8:00	March 3 at 8:00	March 4 at 8:00	March 5 at 8:00	March 6 at 8:00
R155	-39.6	-45.4	-46.6	-47.3	-47.8	-47.3	-48.8
R160	-32.0	-50.6	-56.8	-59.8	-61.3	-60.1	-61.6
R165	-15.7	-62.8	-84.2	-91.4	-92.8	-93.0	-95.2
R170	-2.8	-33.7	-57.1	-67.0	-69.2	-68.5	-79.1

Table 8.6 J01–J06 data of settlement monitoring point of two-floor commercial building (unit: mm)

Measuring point day period	February 28th 8:00	March 1 at 8:00	March 2 at 8:00	March 3 at 8:00	March 4 at 8:00	March 5 at 8:00	March 6 at 8:00
J01	-0.5	-0.4	0.5	0.1	0.6	0.9	1.2
J02	0.1	0.1	0.0	0.0	-1.1	0.2	-1.2
J03	-0.1	-1.8	-2.1	-3.0	-3.7	-3.2	-6.6
J04	0.0	-0.1	-0.7	-1.6	-0.7	-0.4	-2.4
J05	-0.3	-1.3	-1.4	-1.6	-1.4	-1.5	-1.6
J06	-0.1	-0.4	-0.2	0.0	-0.1	-0.3	-0.6

Table 8.7 +164– + 171 Ring shield driving parameters table

Ring plate ginseng The number	Soil bin pressure (bar)	Total thrust (kN)	Knife disk torque (x 10^3 kN/m)	Driving speed (cm/min)	The unearthed amount (m^3)	Synchronous grouting volume (m^3)
164	1.8	13,372	2.6	0.5	39	8
165	1.9	12,394	2.6	0.5	39	8
166	2.0	12,737	2.6	0.5	40	9
167	2.0	15,325	2.7	0.5	40	9
168	2.0	15,737	2.6	1.0	39	9
169	2.0	16,675	2.7	1.0	39	9
170	2.0	15,383	2.7	0.6	38	9
171	2.0	15,470	2.7	0.4	38	9

speed decreases to about 0.5 cm/min, which may cause discontinuity of the tunneling process and increase the disturbance to the surrounding soil. According to observation, the scene of synchronous grouting is limited to adjust the control of grouting quantity and size consistency, content of effective components of slurry choice still is first consider with convenient operation, the scene to avoid grouting pipe blockage occurred frequently, greatly reduce the dosage of fine aggregate, and the increasing dosage of fly ash improve consistency, only make the slurry meet the requirements from the experimental data, but this approach with rich water dumpling–long interval sand layer under the geological don't adapt to the characteristics of the slurry is easily lost in the absence of fine aggregate block under the condition of surrounding rock gap, synchronous grouting, almost no effect for buildings buried under shield the risk.

(3) Risk and reason analysis

1) According to the surface settlement trend analysis of the upper line of the nearby interval in the last week, the surface settlement above the palm during shield tunneling of the upper line has exceeded the control value; During the passage of the shield and after the tube slides out of the tail of the shield, the surface settlement of this position will continue to develop, and most of the subsequent settlement is close to or even may exceed the surface settlement of the upper part of the palm in the early stage. Comprehensive view, long dumpling interval up line in the case of normal excavation of shield machine, surface subsidence rate of each stage and the settlement are far above the control values, single for shield construction engineering already exists a certain risk, and the interval up and down lines are passing under the Phoenix Garden West Building with nearly a week building subsidence monitoring data analysis, we believe that the building settlement and ground surface subsidence will be basic synchronization.

   Combined with the geological conditions of upper line of Jiaochang-long interval and the analysis of the previous construction, the earth bunker pressure of upper line shield tunneling machine has reached 2Bar, which is 1.6 times of the theoretical palm surface earth pressure, which belongs to the strict pressure holding advance, and the possibility of overdigging is not high. Besides, other tunneling parameters are relatively normal, which indicates that there are no major problems in residuum improvement and the adaptability of shield tunneling machine. Speculated that because of its shield is currently under the soft hard formation, the posture is not easy to control, easy to kowtow), and because of ballastless soil flow low plasticity and difficult to realize continuous dynamic soil pressure balance, the overexcavation will occur easily the front surface settlement is too large, some of the settlements will develop to be later settlement. These are the reasons why the surface subsidence rate and subsidence in each stage are still far more than the control value under the condition that the Jiao-Long section uplink shield machine maintains normal tunneling.
2) The upper line shield tunnelling machine has a high equipment failure rate recently. On March 6, the upper line was abnormally shut down due to equipment failure. Subsequently, the average settlement rate of surface settlement monitoring point R170 and building settlement measuring point J03 increased sharply. On analysis was found, after each interval up line stops abnormally surface subsidence monitoring data are sensitively fed back by the rapid increase in subsidence rate. The reason may be in the process of the shield machine downtime, super static pore water pressure dissipation gradually, consolidation

of soil and water and soil stress redistribution caused soil bin pressure gradually decreases, resulting in surface subsidence above the tunnel face. Moreover, the layer that the interval passes through contents large amount of water, this phenomenon is prone to occur, which is the main reason for the frequent sharp increase of interval upward surface subsidence rate.

3) According to the recent surface settlement trend of the upper line of the near jiao–long interval, the upper line shield structure still has a large settlement trend after the tube segments come out of the tail of the shield, sometimes even exceeding the early settlement of the position; Analyzes believed that in the current water rich sand stratum, because the current grouting slurry particle size is less than the surrounding rock of gap size, synchronous grouting slurry yi shun from surrounding rock gap caused by groundwater diaspora synchronous grouting and secondary grouting effect is not good, this is lead to segment the ground settlement after the emergence of shield tail continue to another main reason of rapid development.

(4) Measures and Suggestions

1) All shield parameters should be corrected immediately on the site according to the geological changes to make them adapt to the current stratum conditions. It is suggested that the improvement measures of residuum which is more suitable for enhancing the flow plasticity of residuum in the soil bin should be adopted to further increase the pressure of the soil bin on the basis of the existing soil pressure parameters so as to ensure the shield tunneling smoothly.
2) It is suggested to strengthen equipment support, establish routine overhaul and maintenance system, reduce the abnormal downtime and downtime of shield, reduce soil pressure loss on the palm surface, and avoid excessive settlement of soil above the palm surface and the shield body.
3) In view of the dumpling–long interval slurry high turnover situation, suggest changes of synchronous grouting parameters (appropriate reduce paste increase aggregate content, and encryption of secondary grouting measures), at the same time shorten slurry solidifying time, close-grained filling segment gap with the soil, to ensure that the segment out behind the tail above the soil settlement control within the scope of the permit.

8.4 Case 3: The Tunnel Between Ningbo West Gate and Gulou Station Passes Through Important Cultural Relics

8.4.1 Project Overview

(1) Project overview

West gate—Gulou station is located under Zhongshan West Road, from Xiaowen Street to Jiefang Road, mileage pile number is K9 + 777.0—K10 + 552.0, the total length is about 775 m (double line), the line spacing is 11.9–14.6 m, are all underground lines, with one connecting passage. The route starts from Ximen Station, goes east along Zhongshan West Road, passes Xiaowen Street, Xiantong Tower, Courthouse Lane, Hutong Street, Zhenming Road, Jiefang Road overpass and arrives at Drum Tower Station, with dense buildings on both sides and four important cultural relics below (Xiantong Tower, Fanzhai, Drum Tower and Yongfengku). See Fig. 8.33 for details.

Structural design and construction method: Shield method is used to construct the interval from Ximen Station to Gulou Station. The length is about 632 m and the depth of the structure is 14–23 m. It is a single circular shield method interval. Single layer lining of reinforced concrete with staggered joints; Six blocks were divided into blocks, and K blocks were inserted vertically into the assembly mode. The specific blocks were 3 standard blocks (center Angle 67.5°), 2 adjacent blocks (center Angle 68.75°) and 1 cap block (center Angle 20°).The type of the tube is standard ring + left ring + right ring; The lining ring is 1200 mm wide and 350 mm thick, and the inner diameter of the tunnel is 5.5 m. The lining longitudinal joint section adopts positioning rod and the torus is provided with concave and convex mortise groove. The shield tunneling piece is made of a piece structure connected by bent bolts between blocks and rings. There is a connecting channel and drainage pumping station.

(2) Engineering geology and hydrogeology

West gate–Drum Tower station is located on The West Zhongshan Road, the line is located between Wangjing Road and Jiefang Road. The interval tunnel is generally

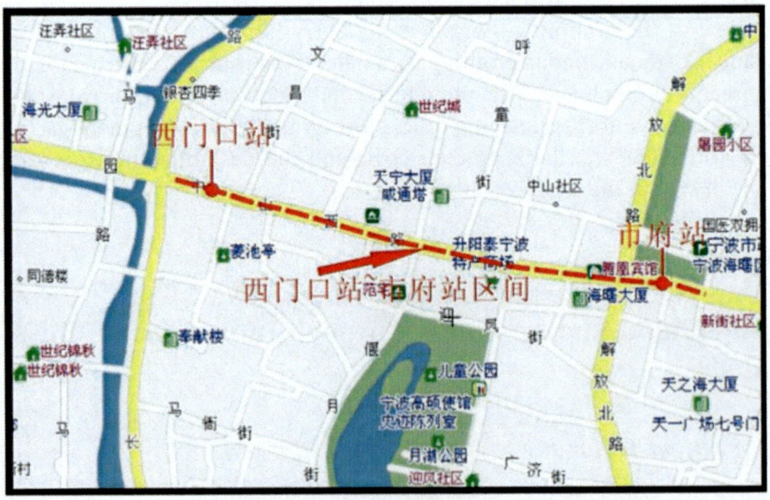

Fig. 8.33 Schematic diagram of plane position of west Gate to Gulou Station

located in ②$_{2-2}$ Layer gray muddy clay, ③$_1$ layer contains cohesive soil silt or argillaceous silt, ③$_2$ Layer silty clay, ④$_{1-1}$ Layer silt silty clay, ④$_2$ Layer clay and ⑤^1 Hard soil layer.

Confined water occurs in in this area ③$_1$ Layers of silt and sand ⑧$_1$. According to regional hydrogeological data, the head buried depth of artesian water is 5.0–7.0 m, which has little adverse impact on shield construction.

(3) Surrounding environment

1) Surrounding structures

The tunnel starts from Ximen Station, goes east along Zhongshan West Road, passes Xiaowen Street, Courthouse Lane, Hutong Street, Zhenming Road and Jiefang Road overpass to Reach The Gulou Station. The two sides of the tunnel are relatively dense, and there are four important cultural relics below (Xiantong Tower, Fanzhai, Gulou and Yongfengku). There are many high-rise buildings on both sides of the tunnel, including Wenchang Hotel, Zhongzhong Building, Tianning Building, Drum Tower Building and Ningbo Municipal Finance Bureau. The surrounding environment of the site is shown in Figs. 8.34, 8.35, 8.36, 8.37, 8.38, 8.39, 8.40, 8.41.

There are many cultural relics passing through the section from West gate to Drum Tower station, including: Xiantong Tower, Fan Zhai, Drum Tower and Yongfeng Ku, among which Fan Zhai and Yongfeng Ku are very close to the tunnel axis.

Fig. 8.34 The high-rise buildings on both sides of Tianning Building and Courthouse Lane

Fig. 8.35 Zhongshi Building on the north of Zhongshan Road

Fig. 8.36 Xiantong Tower

Fig. 8.37 Drum Tower

Fig. 8.38 Yongfengku Ruins

Fig. 8.39 Fan House and Drum Tower Building

Fig. 8.40 Zhongshan Road and high-rise buildings on both sides

Fig. 8.41 Zhenming Road Shopping Mall

① Xiantong Pagoda, built in the fourth year of Xiantong (863) of tang Dynasty, has a history of more than 1100 years. It is the only existing square brick pagoda in Tang Dynasty south of Yangtze River in China. It is now a state key cultural relics protection unit.

The Xiantong tower is a brick structure with a square plane and a floor length of 3.2 m on each side. It covers an area of more than 10 square meters. The top of the outer tower has no tower brake, the whole tower is about 12 m high; The interior space of the tower is square, with the roof of the vault; The tower body is basically straight without obvious inclination, but the erosion and weathering are serious. The tower has five floors, the bottom is slightly higher and has a pot door, the upper tower shrinks layer by layer; The four walls of each floor are equipped with niches for the worship of Buddha statues, surrounded by bricks stacked with an accretion eaves, protruding about 0.7 m.

② Fan residence: established in the late Ming Dynasty Wanli and Tianqi years, 1996 reconstruction, after the establishment of a cultural shopping mall, is now the Ningbo city cultural relics protection units. Fan house covers an area of about 840 m^2, with a building area of 2100 m^2. Taking the north and south as the longitudinal axis and facing south, it is composed of three halls before and after and the left and right auxiliary rooms to form a large scale "Sun" shaped wooden building complex. The main building beam frame adopts the mixed way of carrying and beam type and fighting type. The structure is simple and the material is huge, which should be the typical of the ancient buildings.

③ Drum Tower: Ningbo is the only surviving ancient chenglou ruins, Haishu District cultural relics protection units. Covering an area of more than 700 m^2, the drum tower is more than 8 m high, 16 m deep and 5 m wide. It is a arched stone building. The gate for the three—storey view of the peak. Five studio. The existing tower for the qing Xianfeng five years of reconstruction, the city tower into a three-storey timber structure. In 1930, reinforced concrete watchtowers and alarm clocks were built and mechanical chimes were installed on the towers of the Qing Dynasty. The lower part of the drum towers was a typical Chinese traditional tower style, while the western drum towers in Roman style were built on the towers. Between early 1988 and June 1989, the municipal government carried out major repairs.

④ Yongfengku Site: full name "Ningbo Yuan Dynasty Qingyuan Road Yongfengku site", is the Yuan Dynasty Period of Ningbo yannya storage area site, site is located under 1 m underground, no ground building. The discovery of the site was listed in the "top ten New Archaeological discoveries in China in 2002".In 2006, it was listed as a national key cultural relics protection unit. At present, a park is being built on the site, with the protection and display of the site as the core.

2) Important underground obstacles

There are many buildings along the west gate–Drum Tower station, and the distribution of adjacent pipelines is complex due to the dense and criss-crossing adjacent pipelines. The main underground obstacles in the west gate to Gulou Station are the passageway of court lane and civil air defense project. The passage of Court Lane is the first floor underground, which adopts box structure with a net width of 8.4 m and a net height of 4.2 m. Larsen steel sheet pile is adopted as the enclosure structure. The underground obstacles are mainly old foundations of buildings and underground civil air defense projects. The underground civil air defense project is located in the south of Zhongshan Road, with a buried depth of 0.8–1.5 m, a width of 1.8–3.0 m, and a height of about 2 m.

3) Analysis of key risk factors

According to the analysis of design data and drawings, the risk factors mainly include silty and silty clay layer, interval crossing construction (structure), municipal road, pipeline, shield entrance and exit tunnel, and contact tunnel construction.

According to the geological conditions, environmental conditions and design location conditions, the key risk factors of this work site are further analyzed and identified as Shown in Table 8.8.

(4) Description of construction

The left line shield of The West Drum section was officially launched on January 15, 2012, and the section tunnel was successfully completed on June 11, 2012, advancing the construction for nearly 5 months in total. The right line shield of the section was officially launched on February 10, 2012 and completed on June 18, 2012, promoting the construction for a total of 4 months. The communication channel of the West Drum area has been frozen since June 29, 2012, and the later stage of the melt grouting has been started since September 25, 2012.

Table 8.8 Statistical table of key risk factors analysis

The serial number	Category	Risk factors for	Bad influence
1	Geological and groundwater conditions	Interval crossing ②$_2$Layer silt clay, ③$_1$Layer of sticky silt, ④$_1$Layer silt silty clay, ④$_2$Layer clay, ⑤$_1$Layers of clay	The geological weakness leads to the unstable attitude of the shield, which is difficult to control. The weak link of the tunnel seal is permeated and leaked, and the large surface settlement leads to the deformation of the surrounding environment
2	Environmental conditions	Shield interval above near roads, pipelines, built, and the cultural relics (wenchang hotel, PingQiao village, tianning temple tower, day ning mansion, court lane underground passage, Fan Zhai (cultural relic), drum tower, zheshang bank, drum tower (cultural relic), yongfeng library site (cultural relic), ningbo bureau of finance, fiscal and taxation building), haishu building, silk mansion, hang lung centre, long bridge, etc.)	Shield tunneling causes settlement of important roads, pipelines, cultural relics and structures, which affects traffic safety and building safety
3	Design (stress) position conditions	Shield entry and exit openings	Shield excavation and construction of structural interface cause stress concentration of the structure, and the stress redistribution makes the stress system more complex, resulting in deformation of its own structure and surrounding environment

8.4.2 Monitoring Point Layout

The shield mainly traverses the following cultural relics: Xianfeng Tower (Jc34–Jc37), Fan Zhai (Jc65–Jc80), Drum Tower (Jc130–Jc135), and Yongfeng Repository site (Jc136–Jc145). The time when the shield tunnels went through cultural relics was mainly from March to May, 2012. The monitoring alarm value in this section is based on the design document, with reference to relevant requirements and combined with the actual situation such as the deformation control level of this project, the monitoring and forecast alarm value is shown in Table 8.9.

Table 8.9 Statistical table of monitoring and forecast warning values

The serial number	Monitoring project		The alarm value
1	The surface subsidence		Cumulative variable − 30 mm, change rate 3 mm/d for two consecutive days
2	The surface uplift		Cumulative variable + 10 mm, rate of change 3 mm/d
3	Tunnel settlement, convergence		Cumulative variable 15 mm, rate of change 2 mm/d
4	Building settlement monitoring		Cumulative variable 15 mm, rate of change 2 mm/d
5	Building slope		The accumulative value of inclination reaches 2/1000 or the inclination speed is greater than 0.0001H/d for 3 consecutive days
6	Underground pipeline	Pressure rigid line	The cumulative variable was 10 mm, the rate of change was 2 mm/d, and the relative Angle was 0.01
		Non-pressure rigid line	The cumulative variable was 20 mm, the rate of change was 3 mm/d, and the relative Angle was 0.01
		Flexible pipeline	The cumulative variable was 30 mm, the rate of change was 3 mm/d, and the relative Angle was 0.01

When the monitoring point reaches the alarm value, alarm immediately, analyze the cause and take corresponding measures immediately.

8.4.3 *Analysis of Monitoring Data of Underworn Cultural Relics*

West gate station–between gulou station under wear monitoring statistics in the following table 8.10, in the process of shield tunneling xianfeng tower, king house, drum tower, such as cultural relics are not higher than alarm value, including sinopac library site construction accumulative total value exceeds the alarm, but the wear rate of change of the cultural relics in the process of monitoring tests, the overall deformation is relatively stable, safe and controllable risk.

According to the progress of shield tunneling on site, from April 2, 2012 to April 30, 2012, the underpass of Xianmen Station to Gulou Station went through Courthouse lane. Jc108, the point with the maximum cumulative value, was selected on site for analysis. The cumulative time-history curve of the value is shown in Fig. 8.42.

According to the schedule chart analysis, under the shield during wear, cultural relic building apparently occurred between the sinking, the biggest change rate of -3.34 mm/d, the cumulative value of the largest up to -24.00 mm, accumulative total value and change rate of overweight, under the shield after wearing (May 7, 2012), the building subsidence rate flatten out, accumulative total value more stable as a whole. About -23.7 mm.

According to the progress of shield tunneling on site, from May 24 to June 14, 2012, the upper line of Ximen Station to Gulou Station ran through the Site of Yongfengku. Jc139, the point with the maximum accumulated value, was selected on site for analysis. The cumulative time-history curve of the value is shown in Fig. 8.43.

Table 8.10 Statistical Table of cultural relics monitoring data

A block	Monitoring project	The name of the cultural relics	Maximum cumulative settlement (mm)	Minimum cumulative settlement (mm)	Monitoring date
TJ1102 standard	Settlement of cultural relics	Xianfeng tower	Jc35-1.79	Jc34-0.66	2012/5/24
		King house	Jc58-6.16	Jc55-1.77	2012/4/11
		The drum tower	Jc135-4.08	Jc132-1.29	2012/5/28
		Yongfeng Reservoir site	Jc139-44.2	–	2012/7/4

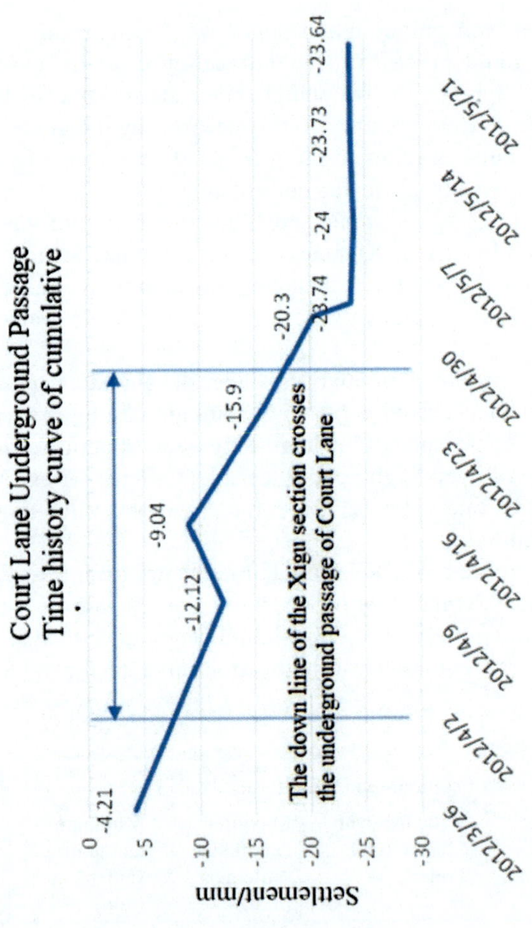

Fig. 8.42 Time history curve of accumulated Jc108 value of underground passage of Court Lane

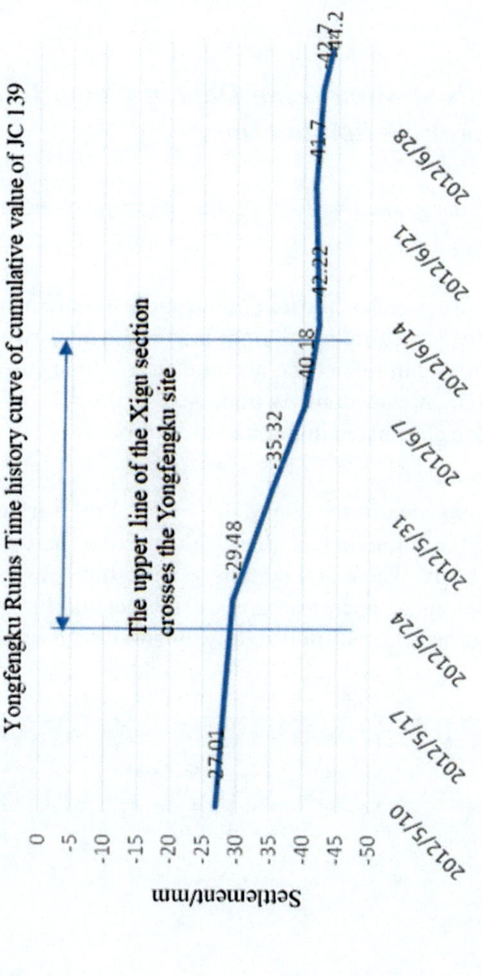

Fig. 8.43 Jc139 cumulative value time history curve of Yongfeng Reservoir site

According to the schedule chart analysis, under the shield during wear, the cultural relic buildings settlement between the cumulative value change significantly, the biggest change rate of 0.81 mm/d, the cumulative value of the largest up to 44.20 mm, considering the previous sinopac library site due to factors such as drum tower station construction settlement to a certain extent, in the process of west drum wear under interval from its cumulative values, under the shield to wear after the completion of the the building subsidence rate flatten out(on June 15, 2012), accumulative total value more stable as a whole. About −44.00 mm.

8.4.4 *Back Analysis of Monitoring Data of Gulou Interval from Ximenkou Station to Line 1*

(1) Cumulative value analysis of vertical section of surface settlement

Selection principle:

1) The measuring point can reflect the maximum surface settlement (located above the center line of the structure on the right line of the interval);
2) The measuring points can reflect the accumulated settlement value caused by the whole process of subway construction;
3) Abnormal points are eliminated in the statistical process of effective measuring points.

Combined with the analysis in the Fig. 8.44, most of the surface subsidence is well controlled during the tunneling of the shield section, and the overall mean value is around −41.73 mm. There is a certain relationship between the over-limit subsidence and the geological characteristics of the section.In the first 550 ring section, the surface deformation control is relatively good, with a maximum value of

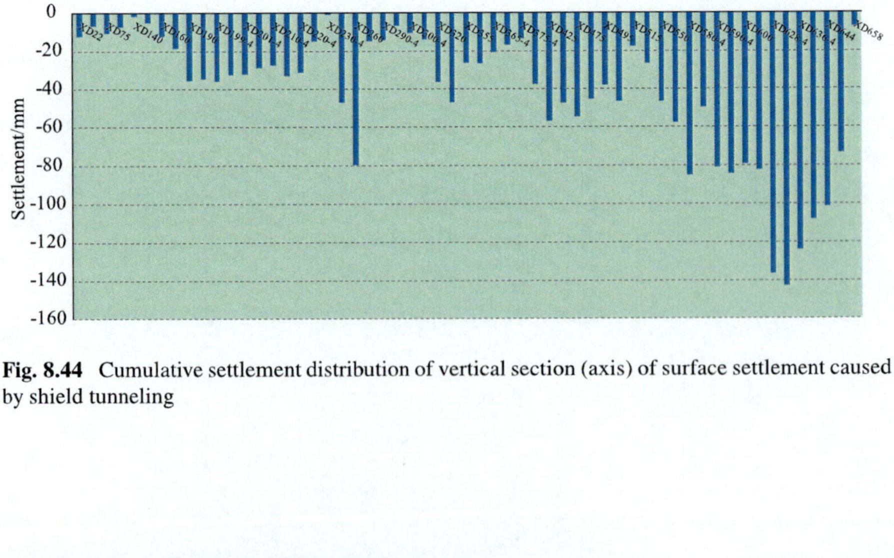

Fig. 8.44 Cumulative settlement distribution of vertical section (axis) of surface settlement caused by shield tunneling

−57.72 mm. However, in the section of XD4-5–XD668 monitored near the end of Gulou Station, the settlement showed an obvious trend of increase, and the surface deformation control was poor.

(2) Cumulative settlement analysis of surface cross-section

32 monitoring sections (cross sections) that have been verified accurately are selected (the data are relatively complete and basically conform to the law of settlement trough for analysis), see Table 8.11.

In combination with the analysis of Figs. 8.45 and 8.46, the overall formation loss control is relatively good in the process of the upper line driving in this area. About 40% sections are less than 10‰ and meet the control requirements of shield driving on the surrounding environment, while a small part of sections (about 13%) meet the control requirements of shield construction on the surrounding environment with the formation loss value between 10‰ and 15‰, which basically meet the control requirements of shield construction. But there are also many section (about 33%) stratum loss value more than 20 ‰, shows that the section caused by factors such as geological shield tunneling control strata loss is too big, there are even a few segments (about 3%) loss rate reached more than 70 ‰, completely unable to meet the requirements of environment protection, which triggered a maximum of strata subsidence is about 142.84 mm. Scope of influence of shield construction: 3.63–35.25 m, mean: 19.44 m.

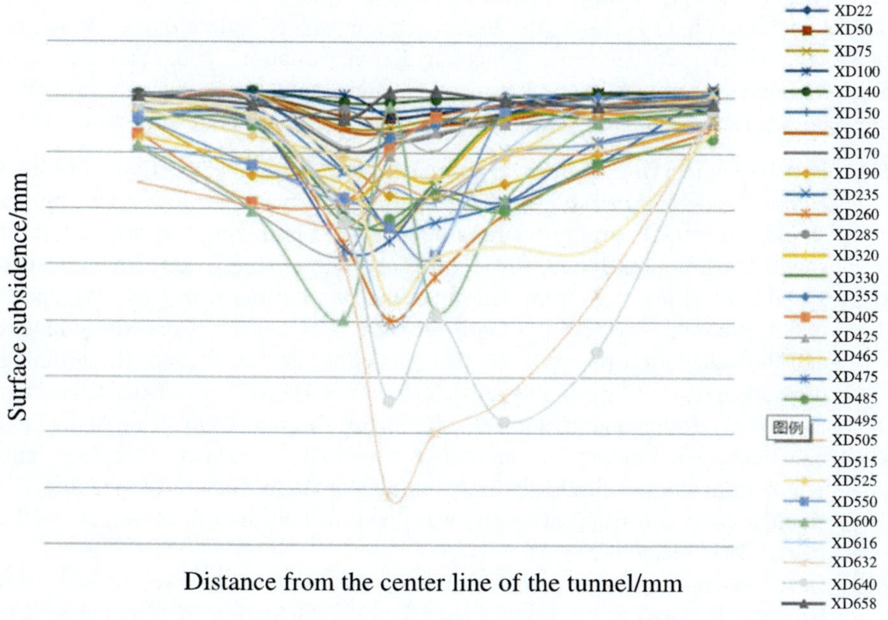

Fig. 8.45 Surface cross-section settlement curve

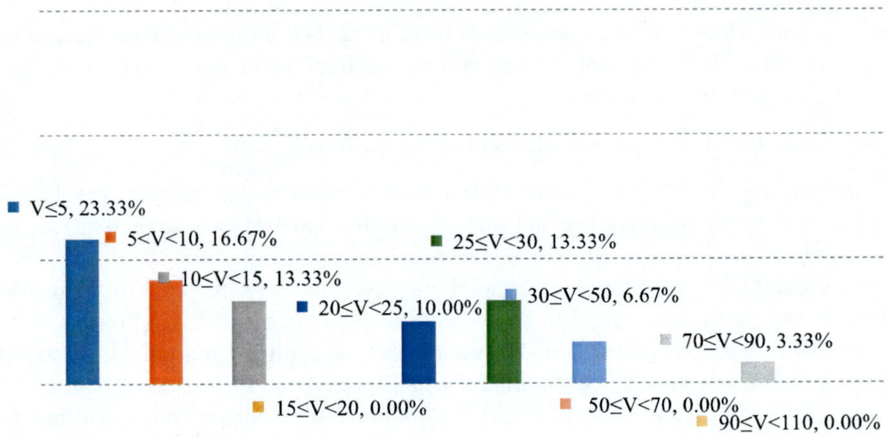

Fig. 8.46 Probability distribution of formation loss rate

8.5 Summary

This chapter studies the implementation effect of shield construction control technology for different adjacent buildings by enumerating three different engineering cases of shield adjacent building construction, including the subway section project undercrossing Fengqi Bridge between "Hangzhou Jianguo North Road Station and Zhonghebei Road Station", the tunnel section project undercrossing west area of Fenghuang Garden between "Nanchang Jiaoqiao Station" and "Yangtze Road Station", and the subway section project undercrossing important cultural relics between "Ningbo Ximenkou station" and "Gulou Station". The results show that:

(1) For the engineering of interval form Jianguo North Road Station to Middle Hebei Road Station (i.e., the knife dish, screw machine, low speed pump, shield tail brush), structure strengthening and optimizing the parameters of shield construction, the construction process to enable speed slow advance, speed control in 5 mm/min, determine the best cutting wheel rotation speed is 0.8 rpm/min. The setting of soil pressure fully considers the safety factor, and the calculation results show that it is controlled at 1.5 bar. The grouting pressure and grouting quantity are determined at 4 bar–5 bar and $4m^3$ according to the early driving parameters and settlement. After the completion of the final construction, the maximum cumulative settlement of the bridge deck was only −3.09 mm, without single alarm or cumulative alarm. Generally speaking, the shield driving construction control was good, without significant impact on the Fengqi Bridge and the river.

(2) Based on the real-time monitoring data feedback of the project of tunnel between Jiaoqiao Station and Changjiang Road Station in Nanchang, it is proposed that the shield structure parameters should be modified according to the geological changes at the construction site, and residue improvement

measures that are suitable for enhancing the flow plastic of residual soil in the soil bin should be selected. Strengthen equipment support, establish routine maintenance system, reduce the number and time of abnormal downtime of shield, reduce the loss of soil pressure on the palm surface, and avoid excessive settlement of soil above the palm surface and the shield body; The synchronous grouting parameters (appropriately reducing the cement-sand ratio and increasing the aggregate content) were modified, and the measures of infilling the secondary grouting were adopted. Meanwhile, the solidification time of the slurry was shortened, and the space between the pipe and the soil was compacted to ensure that the settlement of the soil above the tail of the pipe piece was controlled within the allowable range.

(3) For the important cultural relic project of the tunnel under ningbo Ximeng-Gulou Station, real-time information construction and surface and building deformation should be guided by monitoring and measurement. It can be seen that the change rate of important cultural relic monitoring points in the process of shield tunneling does not exceed the standard, the overall deformation is relatively stable, and the risk is safe and controllable. In addition, most of the surface subsidence control is good, with the overall mean value around −41.73 mm. In the first 550 ring sections, the surface deformation control is relatively good, with a maximum value of −57.72 mm. However, in the XD4-5–XD668 section near the end of Gulou Station, the subsidence volume shows an obvious trend of increase, and the surface deformation control is poor. The overall formation loss control is relatively good, about 40% of the sections are less than 10‰, and a small part (about 13%) of the formation loss value is 10‰–15‰, which basically meets the control requirements of shield construction on the surrounding environment.

It can be seen that different adjacent buildings adopt different control technical measures for shield tunneling, and its selection principle and effect are related to shield tunneling, soil conditions of buildings, characteristics of buildings, relative position of shield metro lines, construction technology and other factors. Therefore, different factors should be taken into consideration in the construction of shield tunneling in the adjacent buildings to select appropriate and reasonable control technical measures, which is of great significance to the protection of buildings and subway structures in the metro tunneling area.